Advanced Processing and Manufacturing Technologies for Structural and Multifunctional Materials III

Advanced Processing and Manufacturing Technologies for Structural and Multifunctional Materials III

A Collection of Papers Presented at the 33rd International Conference on Advanced Ceramics and Composites January 18–23, 2009 Daytona Beach, Florida

Edited by
Tatsuki Ohji
Mrityunjay Singh

Volume Editors
Dileep Singh
Jonathan Salem

A John Wiley & Sons, Inc., Publication

Published by John Wiley & Sons, Inc., Hoboken, New Jersey.
Published simultaneously in Canada.

For general information on our other products and services or for technical support, please contact our Customer Care Department within the United States at (800) 762-2974, outside the United States at (317) 572-3993 or fax (317) 572-4002.

Wiley also publishes its books in a variety of electronic formats. Some content that appears in print may not be available in electronic format. For information about Wiley products, visit our web site at www.wiley.com.

Library of Congress Cataloging-in-Publication Data is available.

ISBN 978-0-470-45758-0

10 9 8 7 6 5 4 3 2 1

Contents

Preface

The Third International Symposium on Advanced Processing and Manufacturing Technologies for Structural and Multifunctional Materials and Systems (APMT) was held during the 33rd International Conference on Advanced Ceramics and Composites, in Daytona Beach, FL, January 18–23, 2008. The aim of this international symposium was to discuss global advances in the research and development of advanced processing and manufacturing technologies for a wide variety of non-oxide and oxide based structural ceramics, particulate and fiber reinforced composites, and multifunctional materials. A total of 68 papers, including invited talks, oral presentations, and posters, were presented from more than 10 countries (USA, Japan, Germany, UK, Ireland, France, Italy, Slovenia, Belgium, Luxembourg, Australia, Brazil, Canada, China, Korea, India, Singapore, Egypt, and Malaysia). The speakers represented universities, industry, and research laboratories.

This issue contains 25 invited and contributed papers, all peer reviewed according to the American Ceramic Society Review Process. The latest developments in processing and manufacturing technologies are covered, including smart processing, advanced composite manufacturing, novel forming and sintering technologies, microwave-processing, polymer-based processing, and film deposition technologies. These papers discuss the most important aspects necessary for understanding and further development of processing and manufacturing of ceramic materials and systems.

The editors wish to extend their gratitude and appreciation to all the authors for their cooperation and contributions, to all the participants and session chairs for their time and efforts, and to all the reviewers for their valuable comments and suggestions. Financial support from the Engineering Ceramic Division and the American Ceramic Society is gratefully acknowledged. Thanks are due to the staff of the meetings and publication departments of the American Ceramic Society for their invaluable assistance.

We hope that this issue will serve as a useful reference for the researchers and technologists working in the field of interested in processing and manufacturing of ceramic materials and systems.

TATSUKI OHJI, *Nagoya, Japan*
MRITYUNJAY SINGH, *Cleveland, USA*

Introduction

The theme of international participation continued at the 33rd International Conference on Advanced Ceramics and Composites (ICACC), with over 1000 attendees from 39 countries. China has become a more significant participant in the program with 15 contributed papers and the presentation of the 2009 Engineering Ceramic Division's Bridge Building Award lecture. The 2009 meeting was organized in conjunction with the Electronics Division and the Nuclear and Environmental Technology Division.

Energy related themes were a mainstay, with symposia on nuclear energy, solid oxide fuel cells, materials for thermal-to-electric energy conversion, and thermal barrier coatings participating along with the traditional themes of armor, mechanical properties, and porous ceramics. Newer themes included nano-structured materials, advanced manufacturing, and bioceramics. Once again the conference included topics ranging from ceramic nanomaterials to structural reliability of ceramic components, demonstrating the linkage between materials science developments at the atomic level and macro-level structural applications. Symposium on Nanostructured Materials and Nanocomposites was held in honor of Prof. Koichi Niihara and recognized the significant contributions made by him. The conference was organized into the following symposia and focused sessions:

Symposium 1	Mechanical Behavior and Performance of Ceramics and Composites
Symposium 2	Advanced Ceramic Coatings for Structural, Environmental, and Functional Applications
Symposium 3	6th International Symposium on Solid Oxide Fuel Cells (SOFC): Materials, Science, and Technology
Symposium 4	Armor Ceramics
Symposium 5	Next Generation Bioceramics
Symposium 6	Key Materials and Technologies for Efficient Direct Thermal-to-Electrical Conversion
Symposium 7	3rd International Symposium on Nanostructured Materials and Nanocomposites: In Honor of Professor Koichi Niihara
Symposium 8	3rd International symposium on Advanced Processing & Manufacturing Technologies (APMT) for Structural & Multifunctional Materials and Systems

Symposium 9	Porous Ceramics: Novel Developments and Applications
Symposium 10	International Symposium on Silicon Carbide and Carbon-Based Materials for Fusion and Advanced Nuclear Energy Applications
Symposium 11	Symposium on Advanced Dielectrics, Piezoelectric, Ferroelectric, and Multiferroic Materials
Focused Session 1	Geopolymers and other Inorganic Polymers
Focused Session 2	Materials for Solid State Lighting
Focused Session 3	Advanced Sensor Technology for High-Temperature Applications
Focused Session 4	Processing and Properties of Nuclear Fuels and Wastes

The conference proceedings compiles peer reviewed papers from the above symposia and focused sessions into 9 issues of the 2009 Ceramic Engineering & Science Proceedings (CESP); Volume 30, Issues 2-10, 2009 as outlined below:

- Mechanical Properties and Performance of Engineering Ceramics and Composites IV, CESP Volume 30, Issue 2 (includes papers from Symp. 1 and FS 1)
- Advanced Ceramic Coatings and Interfaces IV Volume 30, Issue 3 (includes papers from Symp. 2)
- Advances in Solid Oxide Fuel Cells V, CESP Volume 30, Issue 4 (includes papers from Symp. 3)
- Advances in Ceramic Armor V, CESP Volume 30, Issue 5 (includes papers from Symp. 4)
- Advances in Bioceramics and Porous Ceramics II, CESP Volume 30, Issue 6 (includes papers from Symp. 5 and Symp. 9)
- Nanostructured Materials and Nanotechnology III, CESP Volume 30, Issue 7 (includes papers from Symp. 7)
- Advanced Processing and Manufacturing Technologies for Structural and Multifunctional Materials III, CESP Volume 30, Issue 8 (includes papers from Symp. 8)
- Advances in Electronic Ceramics II, CESP Volume 30, Issue 9 (includes papers from Symp. 11, Symp. 6, FS 2 and FS 3)
- Ceramics in Nuclear Applications, CESP Volume 30, Issue 10 (includes papers from Symp. 10 and FS 4)

The organization of the Daytona Beach meeting and the publication of these proceedings were possible thanks to the professional staff of The American Ceramic Society (ACerS) and the tireless dedication of the many members of the ACerS Engineering Ceramics, Nuclear & Environmental Technology and Electronics Divisions. We would especially like to express our sincere thanks to the symposia organizers, session chairs, presenters and conference attendees, for their efforts and enthusiastic participation in the vibrant and cutting-edge conference.

DILEEP SINGH and JONATHAN SALEM
Volume Editors

SOLID-STATE REACTIVE SINTERING OF POLYCRYSTALLINE ND:YAG CERAMIC LASER HOST MATERIALS USING AN 83 GHZ MILLIMETER WAVE SYSTEM

Ralph W. Bruce and Chad A. Stephenson
Physical Sciences Dept., Bethel College
Mishawaka, IN, USA

Arne W. Fliflet and Steven B. Gold
Plasma Physics Division, Naval Research Laboratory
Washington, DC, USA

M Ashraf Imam
Materials Science and Technology Division, Naval Research Laboratory
Washington, DC, USA

ABSTRACT

We are investigating the solid-state reactive sintering of polycrystalline Nd:YAG ceramic laser host materials using a high power millimeter-wave beam as the heat source. The starting powder is a mixture of commercially available alumina, yttria, and neodymia powders. The laser-quality results obtained using the solid-state reactive sintering approach and the same materials in a conventional vacuum furnace[1] provide a benchmark for our experiments, which are being carried out using the Naval Research Laboratory (NRL) 83 GHz Gyrotron Beam Material Processing Facility. One objective of our work is to determine the effect of millimeter-wave heating on processing variables such as temperature and hold time and on the microstructural properties impacting the laser host application. Another objective is to optimize the heating uniformity and efficiency of the process for future use in a manufacturing process. Initial experiments with 1-hour hold times have produced translucent samples whose microstructure is currently being evaluated. Longer processing times (up to 16 hours) were needed to achieve full transparency in a conventional furnace. Hold times longer than 1 hour were also investigated and will be reported.

INTRODUCTION

Single crystal Nd-doped YAG has been the most widely used solid-state laser material[2]. Current materials for solid-state lasers include single crystals of neodymium-doped YAG and neodymium glasses. While single crystals have high thermal conductivity and can operate at high powers, they are costly and limited in size and dopant concentration. Neodymium-containing glasses can be large with reasonable cost but have low thermal conductivity, thereby limiting average power. Significant advantages of transparent polycrystalline laser host materials for high energy laser (HEL) applications, compared to single-crystal materials, are reduced processing temperatures, greatly reduced processing times, elimination of facet and pore structures, and the possibility of higher dopant concentrations. In addition, polycrystalline laser host materials have good thermal conductivity, high mechanical strength, and can be fabricated into large and complex structures.

One of the challenges in developing polycrystalline laser host materials is the need for high quality starting powders doped with the rare earth lasing ion. These are generally not commercially available and often require costly powder preparation techniques. The solid-state reactive sintering of Nd:YAG is of particular interest in this regard because high-purity alumina, yttria, and neodymia powders are commercially available. Moreover, Lee et al. (ref. 1) have shown that laser quality polycrystalline Nd:YAG can be produced by pressure-free solid-state reactive sintering in a conventional vacuum furnace and that the powder preparation requires only low-cost techniques such as ball milling[1].

Millimeter-wave processing has been shown to be an effective alternative to conventional vacuum furnaces for pressure-free sintering of low-loss oxide ceramic materials[3]. It involves direct volumetric heating of the ceramic powder. This often results in superior microstructure with fewer trapped pores, cleaner grain boundaries, and smaller grain size than conventionally sintered materials. These properties are critical to achieving high optical quality, transparent laser host materials. Other advantages of microwave processing include faster heating rates and the capability to sinter at lower temperatures than conventional heating, resulting in a shorter more efficient process[4].

A critical feature of millimeter-wave sintering is stronger coupling to laser host materials than conventional microwaves. Sesquioxides have very low rf loss and do not couple well at low frequencies (*e.g.* 2.45 GHz) compared to high frequencies (*e.g.* 83 GHz). The absorbed power per unit volume in a ceramic is proportional to the microwave frequency ω according to [5]

$$P_{absorbed}(\omega,T)=\frac{1}{2}\omega\varepsilon_0\varepsilon''(\omega,T)|E|^2 \quad (1)$$

where ε_0 is the free space permittivity, ε'' is the relative dielectric loss and E is the local rf field. Thus the power loss is a function of both the temperature T of the ceramic and the frequency; at a given frequency, oxide ceramics tend to be more absorbing at higher temperatures, and at a given temperature, an oxide ceramic is more absorbing at higher frequencies. The frequency dependence of the power absorption is an important motivation for processing low-loss ceramics at 83 GHz rather than at 2.45 GHz or 35 GHz.

We have therefore embarked on an investigation of solid-state reactive sintering of polycrystalline Nd:YAG ceramic laser host materials using an 83 GHz beam as the heat source. We are using the same materials and powder preparation techniques as discussed in ref. 1, so that the results of millimeter-wave processing can be directly compared with conventional vacuum sintering.

Equipment

The NRL gyrotron-based material processing facility (Fig. 1) features a 20 kW, CW (Continuous Wave), 83 GHz GYCOM gyrotron oscillator, which can generate between 1 W/cm^2 and 2 kW/cm^2 irradiance. The facility features a quasi-optical beam system for beam focusing and manipulation and an inner vacuum chamber for atmosphere control and vacuum processing. The gyrotron is operated via a fully automated computerized control system written in the LabVIEW™ platform with feedback from extensive *in-situ* instrumentation and visual process monitoring.

Figure 1. NRL 83 GHz gyrotron-based material processing facility. The millimeter-wave beam furnace is located inside the large processing chamber which serves to confine the millimeter-wave radiation. A 2-color pyrometer looks down on the furnace through a screen at the top of the processing chamber. The millimeter-wave beam exits the gyrotron horizontally (white arrow) and is deflected by a slightly concave mirror into the vacuum furnace through a quartz window.

Figure 2. Millimeter-wave beam furnace. Side view [left], top view with lid including quartz vacuum window removed [right].

Experimental setup and processing

α-Alumina (AKP-50, Sumitomo, Japan), yttria (NYC Co., Tokyo, Japan) and neodymia (NYC Co., Tokyo, Japan) were obtained and mixed in the appropriate ratios to give 0, 1, 2 and 4 at. % Nd.

To this was added approximately 0.5% TEOS (Alfa, Ward Hill, MA) as a sintering aid. Ethyl Alcohol was then added to the mixture which was then ball milled for 16 hours. The milling media was high purity alumina balls. After milling, the slurry was dried and then hand-milled in an alumina crucible into a fine powder.

The green compacts were uniaxially pressed to approximately 53% theoretical density (TD). Some of these were then cold isostatically pressed (CIPed) to densities of approximately 61% TD. The ceramic work pieces were placed in an open or closed crucible and directly exposed to the 83 GHz beam which is focused to a roughly elliptical shape (approximately 1 cm by 4 cm) by the concave mirror. This type of beam is adequate for processing the small compacts currently being tested (diameter 5mm). Larger compacts will require a larger, more uniform beam. The crucible and the materials surrounding the workpiece (casketing) are chosen to provide thermal isolation and low temperature heating, and to reduce radiative losses to the cold-walled furnace. The ceramic work piece may be embedded in a setter powder and/or microwave susceptor. Zirconia is often used as a setter powder and other setter powders include boron nitride, alumina, and yttria. The beam power and intensity at sintering temperatures is a few kilowatts and a few 100 W/cm^2, respectively. The mirror position is adjusted during processing to optimize irradiation of the workpiece. The workpieces are processed in a small vacuum chamber (inner diameter 33 cm, height 28 cm) (Fig. 2) in a vacuum of between 25 and 100 milliTorr. The pressure is monitored for signs of outgassing during initial heating. The temperature is monitored by both an S-type thermocouple (platinum/platinum with 10% rhodium) situated near the sample and a remotely located two-color pyrometer.

The automated temperature controller elevates the sample temperature using feedback from the thermocouple or two-color pyrometer (by increasing the gyrotron voltage and consequently its output power) at a predetermined rate of approximately 10 - 20℃ per minute until it reaches the desire hold temperature. Typical hold time at temperature was approximately 1 hour, though some tests were conducted for a longer period. Final densities ranged from 57% for low-temperature and non-CIP'ed compacts to fully densified compacts. A summary of details is given Table I. A total of 45 samples were processed during this experimental effort.

Table I: Schedule of Processed Samples

83 GHz reactive sintering of Nd:YAG						**$Nd^{3+}:Y_3Al_5O_{12}$**	
		6793-01	**4 at. % Nd**				
Sample	**Temperature**	**Hold Time**	**at.% Nd**	**CIP**	**Density before CIP**	**Density after CIP**	**Final Density**
1	1800	15 min	4%	Yes			94%
2	1800	1 hour	4%	Yes		55%	99%
3	1700	1 hour	4%	Yes		54%	95%
4	1600	1 hour	4%	Yes		60%	90%
5	1900	1 hour	4%	Yes		57%	
6	1600	1 hour	4%	Yes	52%		98%
7	1700	1 hour	4%	Yes	54%	60%	99%
8	1800	1 hour	4%	Yes	54%	60%	93%
10	1800	2 hours	4%	Yes	54%	60%	
11	1750	75 min	4%	Yes	52%		
12	1750		4%	Yes	54%		
14	1750		4%	Yes	54%		

21	1700	1 hour	4%	No	53%		99%
22	1300	1 hour	4%	No	54%		
23	1500	1 hour	4%	No	54%		
24	1500	1 hour	4%	No	53%		
25	1500	1 hour	4%	No	52%		
26	1500	1 hour	4%	No	52%		
27	1500	30 min	4%	No			
28	1000	1 hour	4%	No	52%		59%
29	1100	1 hour	4%	No			68%
30	1500	1 hour	4%	No			74%
31	1200	1 hour	4%	No			61%
32	1300	1 hour	4%	No			62%
33	1400	1 hour	4%	No			57%
6793-02 2 at. % Nd							
Sample	**Tempera-ture**	**Hold Time**	**at.% Nd**	**CIP**	**Density before CIP**	**Density after CIP**	**Final Density**
1	1750	1 hour	2%	Yes		60%	99%
2	1700	1 hour	2%	Yes		61%	98%
3	1800	1 hour	2%	Yes		61%	100%
4	1750	1 hour	2%	Yes			100%
5	1800		2%	Yes			
6	1800	2 hours	2%	Yes			
7	1650	2 hours	2%	Yes			98%
8	1725	2 hours	2%	Yes			98%
6793-03 1 at.% Nd							
Sample	**Tempera-ture**	**Hold Time**	**at.% Nd**	**CIP**	**Density be-fore CIP**	**Density after CIP**	**Final Density**
1	1700	1 hour	1%	Yes		61%	98%
2	1750	1 hour	1%	Yes		63%	
6793-04 0 at. % Nd							
Sample	**Tempera-ture**	**Hold Time**	**at.% Nd**	**CIP**	**Density be-fore CIP**	**Density after CIP**	**Final Density**
1	1000	1 hour	0%	No			59%
2	1100	1 hour	0%	No			59%
3	1200	1 hour	0%	No			61%
4	1300	1 hour	0%	No			65%
5	1400	1 hour	0%	No			72%
6	1500	1 hour	0%	No			90%
7	1700	1 hour		Yes			
8	1750	1 hour		Yes			

Results and Discussion

One of the objectives of this work was to obtain results that would be comparable to obtained in ref. 1. To this end, a number of 0% Nd samples (Samples 6793-04, 1 to 2 and 4 to 6 in Table 1) were sintered after which XRD analysis was performed. The results are presented in Fig. 3, below. Whereas in the comparative study, all yttria and alumina was converted to YAG at a temperature of 1500 °C, there was still some evidence of the perovskite phase (YAP, denoted as P) at our measured temperature of 1500 °C.

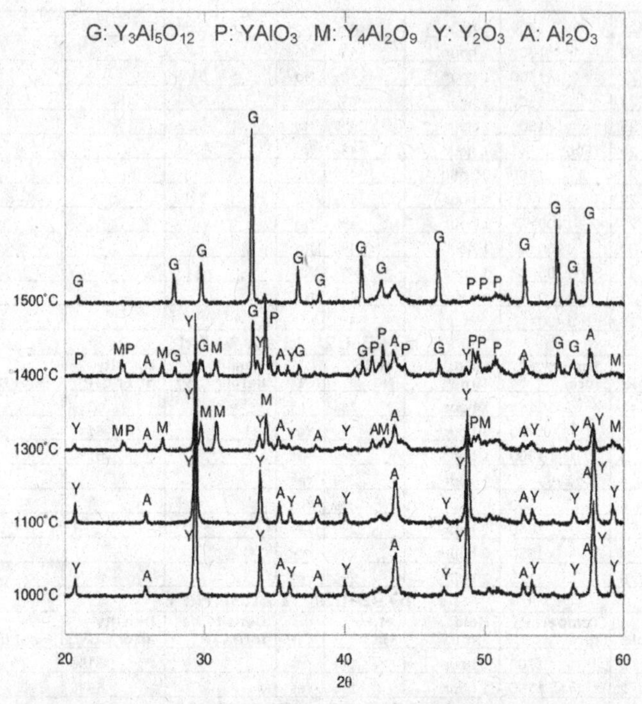

Figure 3. XRD of five samples of the 0 at.% Nd materials [Designated 6793-04 in Table 1]. The legend of the various peaks is given in the figure. Y and A are for the unchanged starting materials. M is the monoclinic phase of YA and begins to appear at 1300 °C or below but is gone by 1500 °C. P is the perovskite phase that begins to appear at 1300 °C and still evident at 1500 °C. The garnet phase, G, begins to appear at 1400 °C. Most of the material is converted to this phase by 1500 °C with a small residual of the perovskite phase.

Another objective of this study was to determine at what temperature full density would be reached. As given in Table 1, this was at approximately 1700 °C. XRD data for Sample 6793-02 (2 at. % Nd) is given in Figure 4, below. The garnet phase, G, is the only phase present at this temperature as was to be expected from the previous study. This is agreement with the results obtained in ref. 1. Yet it should be noted at this time, our results were usually obtained in 1 hour of hold time as contrasted to the 16 hours required by the reference study. Holds at the longer period of 2 hours did not show any significant gain in terms of increased densification.

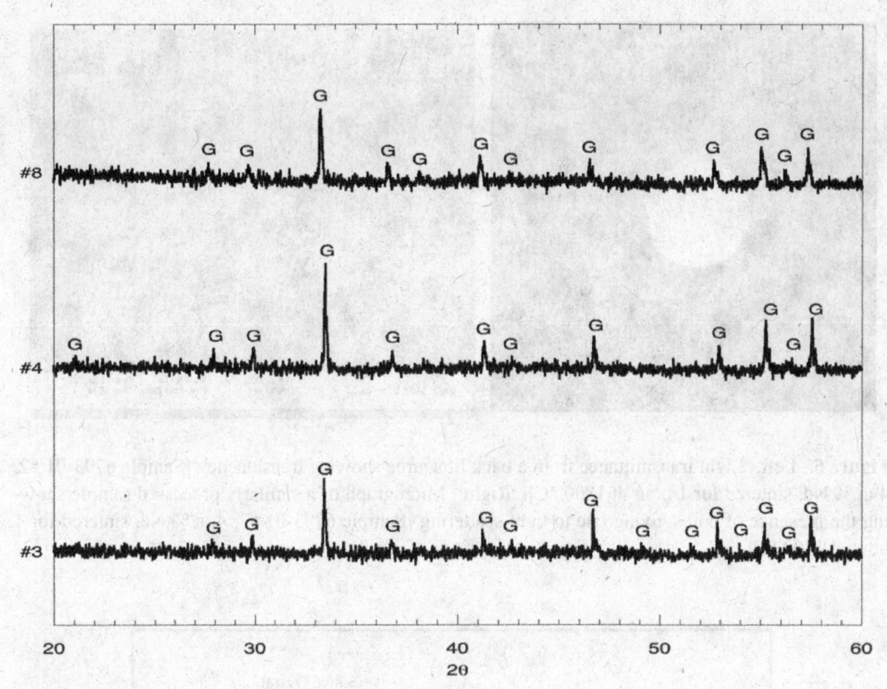

Figure 4. XRD of 2 at. % Nd:YAG. #3 was processed to 1800 °C with a hold time of 1 hour. #4 was processed to 1750 °C and #8 to 1725 °C with hold times of 1 hour.

Fundamental to this study was the obtaining of transparent materials. This was not accomplished to the degree desired but a high of translucency was in fact obtained. Figure 5 (left) shows an example of one of the more successful examples. This sample (6793-02 #4) is displayed with a light source from behind. Transmittance analysis was not performed. Analysis of the cause of the translucency was performed by looking at the microstructure using an SEM. This was performed on a sample similarly processed as was the foregoing sample. The result is shown in Figure 5 (right). The existence of pores gives rise to light scattering and a resultant non-transparent compact. It should also be noted that the grain sizes for our samples are on the order of 10 μm or less. The grain sizes for the reference study were on the order of 50 μm.

Notwithstanding the lack of transparency of the sample, the most important consideration in working with laser host materials is whether or not the Nd-doped material will in fact behave as a laser material. One critical indication of this is the measurement of the fluorescence lifetime once the material has been pulsed with the appropriate light source. Figure 6 shows the values we obtained for the 6793-02 (2 at. % Nd) samples. They compare very favorably with the results from other studies[6] In particular, the reported values for polycrystalline Nd:YAG are 184 μs whereas the NRL material had an average value of 197 μs with 203 μs for the 1725 °C sample to 188 μs for the 1800 °C sample. This temperature dependency may or may not be significant and will be examined in future studies.

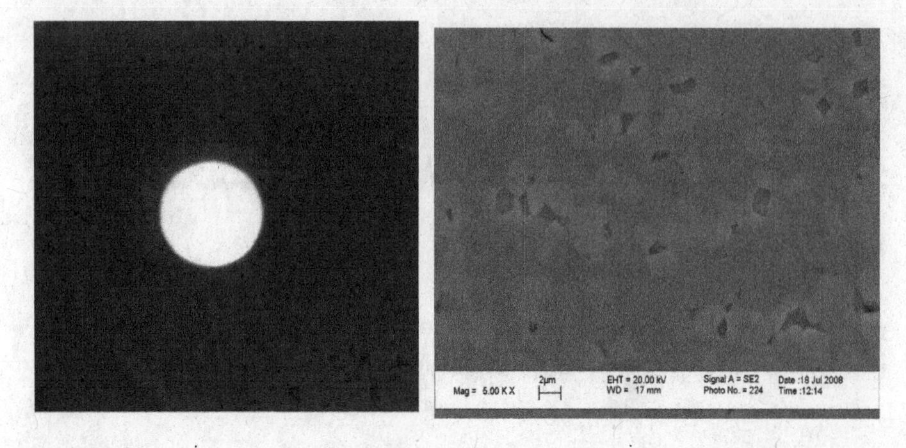

Figure 5. Left: Light transmittance from a back lit source showing translucency [Sample 6793-01 #2, 4 at.% Nd, sintered for 1 hour at 1800 °C]. Right: Micrograph of a similarly processed sample showing the presence of pores giving rise to light scattering [Sample 6793-01 #3, 4 at.% Nd, sintered for 1 hour at 1700 °C].

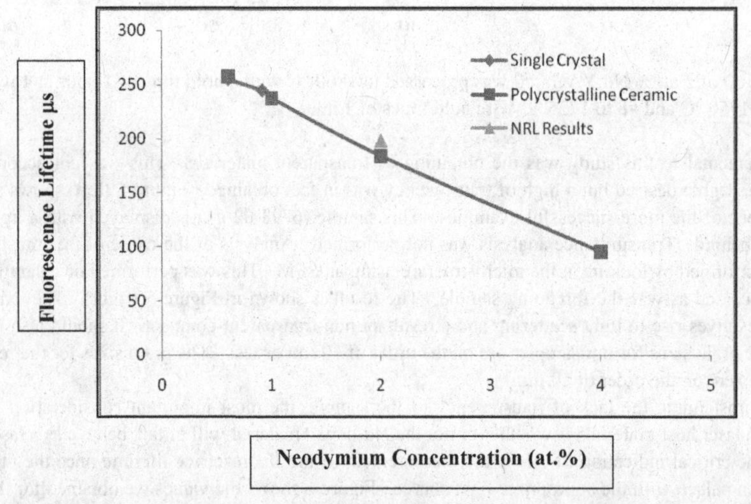

Figure 6. Fluorescence lifetime measurement of the 6397-02 samples 3, 4 and 8 compared to published results (ref. 6).

Conclusion

Translucent polycrystalline Nd:YAG samples , having fluorescence lifetimes consistent with published results, have been produced using the high-powered 83 GHz Millimeter-Wave Processing Facility at the Naval Research Laboratory. These early results indicate that this method of reactive sintering can produce viable material in significantly less time when compared to conventional ceramic processing techniques.

Acknowledgement

This work is sponsored by the Office of Naval Research including the ONR/ASEE Summer Faculty Fellowship Program and was supported by the Directed Energy Professional Society through a Directed Energy Scholars Grant to C. A. Stephenson.

The authors would also like to thank Dr. Feng of the Materials Science and Technology Division for his assistance in the providing x-ray and SEM data.

References

[1] S.-H. Lee, S. Kochawattana, G. L. Messing, J. Q. Dumm, G. Quarles, and V. Castillo, Solid-State Reactive Sintering of Transparent Polycrystalline Nd:YAG Ceramics, *J. Am. Ceram. Soc.* **89** [6] 1945-1950 (2006).

[2] A. Ikesue and I. Furusato, Fabrication of polycrystalline, transparent YAG ceramics by a solid-state reaction method, *J. Am. Ceram. Soc.*, **78** 225-28 (1995).

[3] D. Lewis, M. A. Imam, *et al.*, Material processing with a high frequency millimeter-wave source, Mat. Manuf. Proc. **18,** 151-167 (2003).

[4] R.W. Bruce, *et al.*, Joining of ceramic tubes using a high-power 83-GHz Millimeter-wave beam, IEEE Trans. Plasma Sci. **33**, 668-678, 2005

[5] K. G. Ayappa, H. T. Davis, E. A. Davis, and J. Gordon, Analysis of Microwave Heating of Materials with Temperature-Dependent Properties, AIChE Journal **37**, 313-322, 1991.

[6] J. Lu, M. Prabhu, J. Song, C. Li, J. Xu, K. Ueda, A.A. Kaminskii, H. Yagi, and T. Yanagitani, Optical properties and highly efficient laser oscillation of Nd:YAG ceramics, *Appl, Phys. B*, **71**, 469 – 473 (2000), DOI:10.1007/s003400000394

MICROWAVE ASSISTED LARGE SCALE SINTERING OF MULTILAYER ELECTROCERAMIC DEVICES

B.Vaidhyanathan[1,*], K.Annapoorani[1], J.G.P.Binner[1] and R.Raghavendra[2]
[1]Department of Materials, Loughborough University, Leicestershire, LE11 3TU, UK
[2]Littelfuse Ireland Ltd, Dundalk, Ireland

ABSTRACT

The feasibility of employing the microwave methodology for the processing of integrated passive devices (IPDs), nanocrystalline ZnO radials and nano multilayer varistor (MLVs) devices was explored. Methodical microwave sintering experiments were carried out using a multimode, 2.45 GHz microwave applicator. Effect of various experimental parameters such as heating rate, cooling rate, soaking time, sintering temperature etc. on the processing of these device components was investigated in detail. The resultant products were characterized for microstructure, composition and electrical performance. The various stages involved in taking the laboratory research to industrial scale-up production were also examined. The use of microwaves for the processing of MLVs was found to genuinely improve the electrical properties in both small scale (~200 devices/ batch) and large scale (~12000 devices/batch) sintering situations. For a stand alone microwave heating process a back-to-back cascading /conveyer belt arrangement is recommended for continuous large scale production. However hybrid heating methodology was found to provide the capability of stacking operations and could be helpful in avoiding the use of 'casketing', besides providing the possibility of achieving uniform temperature across a large volume. The technique seems to be attractive in terms of its simplicity, rapidity, economic viability and the superior product performance achieved in all the cases augers well for its general applicability.

INTRODUCTION

Electroceramic devices such as varistors, capacitors, transducers and integrated devices constitute a multibillion-dollar global market owing to their applications in most modern electronic appliances. Each of these devices consists of a dielectric layer sandwiched between metal electrodes. The key step that determines the electrical properties of such a device is sintering, which converts the green component into a dense monolithic device. The conventional manufacturing procedures available for the fabrication of these devices involve tedious multi-step binder removal stages, long processing durations and high sintering temperatures. These cumbersome procedures lead to low equipment productivity, unwanted diffusion of electrode (eg., in most multilayer structures) into the ceramic, excessive grain growth (in the case of nanocrystalline varistors), ceramic inter-diffusion at the component interface (in the case of integrated devices) and in turn inferior electrical performance. This necessitates the need to look for alternate and improved processing methodologies. It has been demonstrated by us and research laboratories around the globe that microwave heating is a viable method for the laboratory scale processing of electroceramic devices such as varistors, capacitors, integrated devices etc, owing to the associated advantages viz., savings in time and energy, improved properties and retaining of finer microstructures.[1-7] However the commercial application of microwaves in high temperature sintering processes for advanced ceramics has been slow especially considering these tremendous opportunities.[8,9] The barriers to commercialisation have included lack of microwave furnace suppliers, lack of understanding of the material-microwave interactions, the need for quick, convincing proof of concept studies prior to investment in equipment and the need for reproducible and scale-up process development. The present work describes the salient results obtained on the use of microwaves for the processing of some of the electroceramic devices and we envisaged that the inherent advantages of rapidity, inside-out heating and faster sintering kinetics can be exploited to circumvent some of the processing problems normally encountered during multilayer device manufacture.

EXPERIMENTAL

The microwave cavity used is capable of providing a maximum output power of 6 kW and operating at 2.45 GHz[10]. The microwaves from the magnetron were transmitted through a waveguide system and launched into the cavity after reflecting onto a mode stirrer rotating at 70 rpm. The cavity was provided with a water-cooled base plate to hold the samples and an outer cooling jacket for safety. A reflected power monitoring unit and a three-stub tuner helped to tune the microwave system with minimum reflected power. Platinum sheathed, shielded R-type thermocouple and an infrared pyrometer (LAND instruments, UK) were used for temperature measurements[11,4,5]. Fiberfrax high purity insulation was used as a casket for heat containment. In the second part of the work, we have investigated the potential for using hybrid heating for the scale up sintering of MLVs so as to avoid the requirement of using a casket system (thereby avoiding the handling problems in industrial production units) and b) to optimise the amount of microwave power required to achieve the desired property improvements. The hybrid sintering cavity used is described elsewhere[12]. It also operates at 2.45 GHz frequency and is capable of providing a maximum microwave output power of 1100 W and a conventional power of 5000 W. Comparative conventional heating experiments were also performed using an electric furnace. Density measurements were performed on sintered samples using the Archimedes method using deionised water. The samples were investigated using optical microscopy (Olympus Microscopes, Japan) for electrode delamination and interface cracking. Scanning electron microscopic measurements were carried out using SEM; Leica Cambridge Stereoscan S360 and the extent of diffusion of matrix components studied using energy dispersive X-ray analysis (EDX). The electrical properties of the microwave sintered MLV samples were measured at Littlefuse Ireland Ltd. I-V characteristics from 0.1μA·to 0.01A was measured using a Keithley instrument (Model 2410, USA) and from 0.1A to 100A was measured by using a Keytek Model 711, USA surge generator. Around 100 representative samples for each experimental condition were examined for electrical performance. The electrical results were used as a pointer for optimising the experimental parameters of the subsequent microwave assisted sintering trials.

RESULTS AND DISCUSSION

1. Microwave Processing of Integrated Passive devices

In the modern electronic industry, the drive for component counts reduction per board and miniaturization of electronic circuits made the use of integrated components a necessity. The integration of multiple passive components into a single integrated passive device (IPD) offers significant benefits in terms of space reduction and assembly cost. However inter-diffusion of matrix components at the ceramic interface is the problem of great concern during the conventional firing of these devices. In the present case, IPD samples with a capacitor/varistor formulation were considered for investigation. The capacitor formulation is bismuth niobate doped with magnesium oxide and the varistor formulation consists of 98% of zinc oxide and 2% addition of antimony oxide, bismuth oxide, cobalt oxide, etc. Microwave sintering experiments were performed between 950°-1100°C and the soaking time varied between 20-60 minutes. From Figure 1, it is clear that at any given temperature the

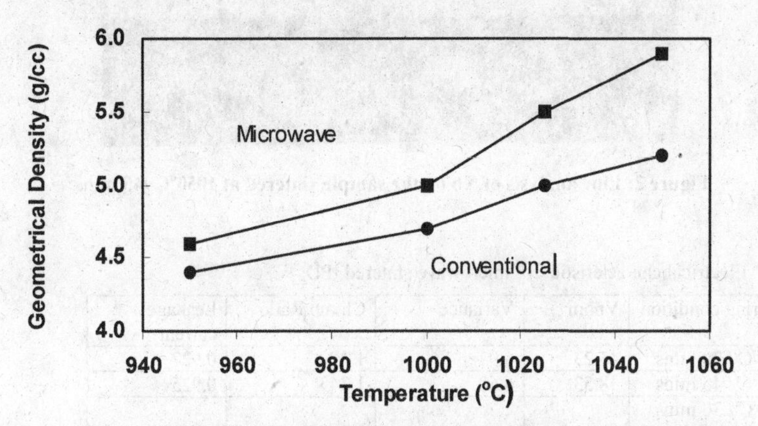

Figure 1: Densification of IPDs in microwave and conventional sintering

microwave process provides much higher densification than conventional sintering and at higher temperatures (>1050°C), the capacitor region sintered more than the varistor part, leading to non-uniform shrinkage and device de-lamination.

The microwave sintered samples were further characterised using EDX and from figure 2 showing the Nb line analysis, it is clear that there is no diffusion of Nb at the ceramic interface (which is the major problem during the long duration conventional processing). Since the Nb lines overlap with that of the electrode material, sharp peaks are seen at the electrode positions (the sharpness of these peaks indicate the well-preserved nature of the electrodes). In contrast, it was observed that in the case of conventionally co-fired composite MLCC's there is an appreciable amount of electrode material migration into the ceramic layers[13]. In this context, the microwave sintering appears to be very attractive since it not only minimises the inter-diffusion of ceramic components but also reduces the electrode migration. Importantly using microwaves, IPDs could be sintered to more than 98%

densification within 3 hours of total cycle time, which is one tenth of the time required by conventional methods[14].

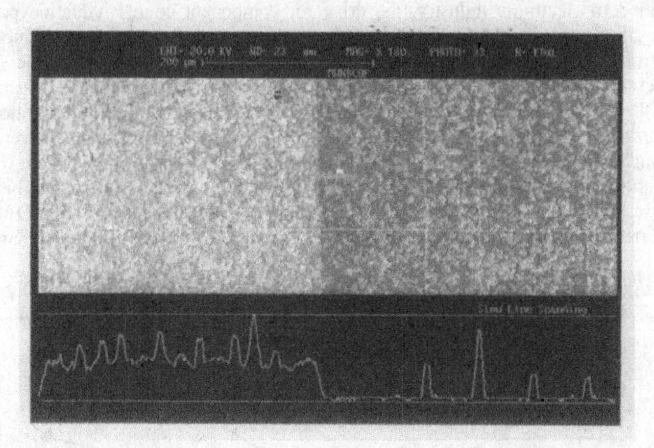

Figure 2: Line analysis of Nb on the sample sintered at 1050°C /45 mins

Table I. Electrical characteristic** of microwave sintered IPDs

Sintering condition	Vnom	Variance	Clamp Ratio	Leakage current
1050°C / 30 mins	45.23	6.93	1.290	0.927
1050°C / 45 mins	38.53	6.60	1.343	0.925
1050°C / 30 mins (flowing air + better casket)	43.57	3.93	1.323	0.930
Required values	>25	~3.5	<1.5	>0.8

Table I provide the electrical performance of microwave sintered IPDs at 1050°C with various soaking times. It is clear that the electrical performance of the microwave sintered IPDs were found to match, and in most cases better, the required values. However initially the parameter 'Variance', was found to be higher, was circumvented by adopting a better casket design for heat containment and passing a burst of air to achieve thermal equilibrium during the microwave processing of these multilayer devices. Use of secondary susceptors was also found to be beneficial.

2. Microwave processing of Nanocrystalline Varistor Devices

In this case, the major objective is to reduce the processing temperature (so that high Ag containing electrodes can be used) of the varistor devices by using the combination of nanocrystalline formulations and microwave methodology. Initial microwave sintering experiments were performed on nano varistor radials. Commercial varistor samples were also microwave sintered for comparison. Figure 3 shows the electrical characteristics of the microwave sintered varistors (at 1100°C/15 mins) and it is evident that the nano sample exhibits excellent electrical performance, with the nominal voltage (Vnom) increased significantly (from 650 V to >1100 V at 1 mA current; the actual value is higher than this but due to equipment limitation we couldn't measure larger values – see the flatness of the voltage due to equipment saturation. Therefore another equipment is used for measurements involving currents higher than 1 mA). To reduce the processing temperature, radials made from another varistor formulation (obtained from Shanxi Corporation, China) were investigated. Using a novel microwave assisted two stage sintering methodology[12] (where in the difference in kinetics between the grain boundary diffusion and grain boundary migration is exploited[15] to minimise the grain growth whilst achieving the required densification) nano varistor devices have been sintered at temperatures lower than 1000°C.

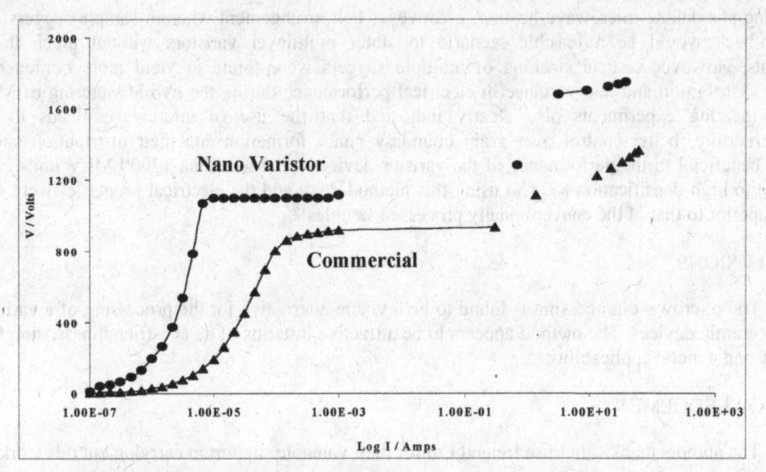

Figure 3: Electrical characteristics of the microwave sintered nano and commercial varistor samples

Multilayer structures based on nano varistor formulation were fabricated with Ag and Ag/Pd electrodes and were microwave sintered successfully to more than 98% densification at temperatures lower than 950°C. The electrical performance of these devices were found to be superior to that of micron sized commercial counter parts and the characteristics were on par with expensive Pt electroded devices[16,17].

3. Scale-up production of Multilayer devices

Successful scale-up sintering of multilayer varistors (MLVs) was carried out by employing both standalone microwave and hybrid heating techniques. When we used a single sagger and ~2000-4000 MLV devices in a larger 6 kW microwave cavity for the initial scale up sintering experiments it was found that at heating rates >25°C/mins, the large, dense ZnO liners commonly used in industrial practice to hold the dielectric samples were found to crack. However by performing a number of methodical experimental trials, this was found to be avoided by employing optimum sintering schedules viz., heating time: 90-100 minutes; soaking time: 20-30 minutes; cooling was achieved by simply switching-off the magnetron and the total cycle time was between 150-180 minutes[18]. All the samples sintered well to obtain >99% densification. In contrast, the conventional sintering schedule using an electric furnace involves a 3 h soak at 350°C and 2h soak at 550°C for complete binder removal and a 12 h soaking time at the sintering temperature of 1200°C resulting in a 30 h total cycle time to achieve >98% densification. Thus almost 90% reduction in processing time was achieved using the microwave methodology whilst obtaining the similar densification. Also the microwave procedure does not necessitate a separate binder removal step, which is a major processing advantage. A number of further sintering experiments were performed involving multiple saggers (3-4) aligned either vertically or horizontally to increase the through-put. Based on the results on density distribution, temperature uniformity and electrical performance[18] it was suggested that in practical situations involving standalone microwave heating, a conveyer belt arrangement wherein sample saggers come back-to-back would be a feasible scenario to sinter multilayer varistors without much thermal gradients. However vertical stacking of multiple saggers were found to yield more homogeneous density distribution and less variance in electrical performance during the hybrid sintering of MLVs. Hybrid heating experiments also clearly indicated that the use of microwaves leads to finer microstrucrture, better control over grain boundary phase formation and their distribution and are indeed beneficial to the performance of the varistor devices[16,18]. More than 12000 MLV units can be sintered to high densification (>98%) using this methodology and the electrical properties were found to be superior to that of the conventionally processed samples[18].

CONCLUSIONS

The microwave processing is found to be a viable alternative for the processing of a variety of electroceramic devices. The method appears to be attractive in terms of its eco-friendliness, simplicity, rapidity and general applicability.

ACKNOWLEDGEMENTS

The authors thank Littelfuse Ireland Ltd for their valuable support in carrying out this work.

REFERENCES

1. R.J. Lauf, C.E. Holcombe and C. Hamby, Microwave sintering of multilayer ceramic capacitors, *Materials Research Symposium Proceedings*, **269**, 223-229 (Materials Research Society, 1992).
2. L.M. Levinson, H.A. Comanzo and W.N. Schultz, Microwave sintering of ZnO varistor ceramics, *Materials Research Symposium Proceedings*, **269**, 311-321(Materials Research Society, 1992)
3. B. Vaidhyanathan, R. Raghavendra, and D.K. Agrawal, "Firing of electrodes for the production of passive devices", *European Patent,* EP 1100095 (2001).
4. B. Vaidhyanathan, J. Wang, J.G.P. Binner and R. Raghavendra, The effect of conventional, microwave and hybrid heating on the sintering of ceamics, *Proc. of the 9th AMPERE conference on microwave and RF heating*, edited by J. Binner, Loughborough, United Kingdom, 31-34 (2003).
5. D.K. Agrawal, R. Raghavendra and B. Vaidhyanathan, Production of Passive Devices, *US Patent*, 6,399,012 (2002).
6. T.R. Shrout, D.K. Agrawal and B. Vaidhyanathan, Microwave sintering of multilayer dielectrics with base metal electrodes,*US Patent*, 6,610,241 (2003).

7. J.G.P. Binner and B. Vaidhyanathan, Microwave Sintering of Ceramics: What does it offer?, *Key Eng. Mater.*, **264-268**, 725-729 (2004).
8. H.S. Shulman, Microwaves in High-Temperature Processes, Industrial Heating Magazine: *The Int. J. Therm. Tech.*, 43-47 (March 2003).
9. R.F. Schiffman, Commercializing microwave systems: Paths to success or failure, *Ceram Tran.*, **59**, 7-14 (1995).
10. N.A. Hassine, J.G.P. Binner and T.E. Cross, Synthesis of Refractory Metal Carbides via Microwave Carbothermal Reduction, *Int. J. Refractory Metals and Hard Mat.*, **13[6]**, 353-358 (1995).
11. B. Vaidhyanathan and J.G.P. Binner, Microwave Assisted Synthesis of Nanocrystalline YAG, *Proc. of the 9th AMPERE conference on microwave and RF heating*, edited by Jon Binner, Loughborough, United Kingdom, 273-276 (2003).
12. J.G.P. Binner and B. Vaidhyanathan, Processing of bulk nanostructured ceramics, *J. Eur. Ceram. Soc.* **28**, 1329-1339 (2008).
13. R. Zue, L. Li, Y. Tang, and Z. Gui, Mutual diffusion at heterogeneous interfaces in co-fired composite MLCCs, *British Ceram. Trans.*, **100**[1], 38-40 (2001).
14. B. Vaidhyanathan, K. Annapoorani, J.G.P. Binner, and R. Raghavendra, Microwave Sintering of Integrated Passive Devices, *Microwave and Radio Frequency Applications,* Edited by D. C. Folz, J.H. Booske, D.E. Clark, and J.F. Gerling, 423-428 (2003).
15. W.I. Chen, X.H. Wang, Sintering dense nanocrystalline ceramics without final-stage grain growth, *Nature*, **404**, 168-171 (2000).
16. K. Annapoorani, "Microwave Processing of electroceramic devices", *Ph.D Thesis*, Loughborough University, UK (2005).
17. K. Annapoorani, B. Vaidhyanathan, J.G.P. Binner and R. Raghavendra, Microwave Assisted Processing of Nanocrystalline Varistor Devices, *J. Euro. Ceram. Soc.*, To be Submitted (2009).
18. B. Vaidhyanathan, K. Annapoorani, J.G.P. Binner, R. Raghavendra and D.K. Agrawal, Microwave Assisted Large Scale Sintering of Multilayer Varistor Devices, *J. Mater. Sci.*, Submitted (2009).

---

* Author for correspondence: B.Vaidhyanathan@lboro.ac.uk ; PH: +44 1509 223152; FAX: +44 1509 223949

** Nominal voltage (Vnom): This is the voltage measured corresponding to 1mA current (V_{1mA}) through the varistor.
Clamp ratio (C_r): This parameter gives the ability of the varistor and is expressed by $C_r = V_c / V_{1mA}$ where Vc is the voltage measured for a given peak current Ic.
Leakage: It is the maximum current (or corresponding voltage) at which the varistor can be operated without getting significantly heated. Beyond this limiting current the joule heating (I^2R) of the ceramic becomes dominant and leads to degradation.
Variance: It is the square root of the standard deviation of the Vnom values.

INFLUENCE OF THE SECONDARY PHASE COMPOSITION ON THE MICROWAVE SINTERING PROCESS

Masaki Yasuoka[1)], Takashi Shirai[2)] and Koji Watari[1)]

1) National Institute of Advanced Industrial Science and Technology (AIST)
2266 Anagahora, Shimo-Shidami, Moriyama-ku, Nagoya, Aichi 463-8560, Japan
2) Ceramics Research Laboratory, Nagoya Institute of Technology
Asahigaoka 10-6-29, Tajimi, Gifu 507-0071, Japan
e-mail address: yasuoka-m@aist.go.jp

ABSTRACT

The ceramic industry uses enormous amounts of energy, in order to make products at high temperatures. In order to protect the earth's environment, the manufacturing industry has to reduce their energy consumption. Energy saving measures based on sintering process improvements are studied. The microwave sintering is expected to be an effective processing method that saves energy. The advantages of microwave sintering are as follows: 1) self-generation of heat, 2) selective heating is possible, and 3) rapid temperature rise and drop. If these features can be employed efficiently, it is thought that not only a process that is environmentally friendly is developed but also materials with the new characteristic can be obtained.

Barium titanate in the presence of a liquid phase was irradiated by a continuous use of lower, increasing levels of power (power-control method). We conducted experiments by varying the amount and method of barium borate. As a result, it has been understood that the samples were densified at temperatures below those at which the liquid-phase is normally formed which reported by phase diagram. In a wide temperature region, the sample made by using the reaction sintering was densified than the sample that made the barium borate beforehand.

INTRODUCTION

To concern about the environment of the Earth, The manufacturing industries for the 21st century will have to reduce their consumption of energy. On the contrary, the ceramic industry uses many heating processes to make those products with high temperatures as necessary and the length of heating requires a sizeable amount of energy. Ceramics are generally recognized to be sintered, when these are conditioned on the uniform temperature distribution in the gas or the electric furnace. This process would be able to apply well for mass production. However, a large amount of energy is consumed due to maintaining the temperature of the inside furnace material or the container, rather than being used in the manufacturing for the target product. If the energy shall be used more efficiently to concern of energy conservation, the sintering process should be done by less energy, also. Consequently, the microwave sintering process has been attracting attention since the 1990s [1,2)] as being energy efficient.

Barium titanate is widely used as a material for multilayer ceramics and thick-film capacitors because of its high dielectric constant [3,4)]. Because barium titanate is used in large quantities as a material in electronic devices, considerable energy savings could be achieved if barium titanate could be sintered at a reduced temperature. It is reported that for LTCC (Low Temperature Co-fired Ceramics) application, barium titanate is sintering at 1000 ℃ for 24 hours with the incorporation of silicate glass system [5)]. In addition, there is a report saying that a sintered body of relative density around 90% was achieved by 900 ℃ for 8 hours by adding boron oxide or lead borate to a barium titanate [6)]. Our aim in this research was to develop a low-temperature sintering method that combines the microwave sintering method, which is expected to give an energy-saving effect, with the liquid-phase sintering method currently used as a low-temperature sintering method [7–11]. The liquid-phase component used was barium borate, which forms a liquid phase was melting at 924 ℃ when mixed with barium titanate [12]: no other chemical components are formed. We conducted experiments by varying the amount of barium borate and modifying its method of addition. We examined the addition method of the fluid phase to make a sample by a simpler method. Also, we examined the effects of combining microwave sintering and liquid-phase sintering and we compared the resulting samples with samples produced by conventional heating in an electric furnace.

EXPERIMENTAL

Barium titanate (BT-01, Sakai Chemical Industry Co., Japan) was used as a raw material. Barium borate, used as the liquid-phase material, was prepared in situ from barium carbonate (99.9%, Wako pure chemical industries Co., Japan) and boric acid (99.5% (minimum), Kanto Kagaku, Japan). The amount of liquid-phase material added was varied between 5, 10 and 20 mol%. The addition was carried out by two methods; 1) barium carbonate and boric acid were mixed with barium titanate; we call reaction sintering method, and 2) barium carbonate and boric acid were mixed beforehand and calcined at 1120 °C to form barium borate and then this was mixed with barium titanate; we call premixed method.

Green pellets were formed by using a uniaxial pressure of 17 MPa and a cold isostatic pressure of 98 MPa. A magnetron multimode microwave furnace (MW-Master, Mino Ceramic Co. Ltd., Mizunami Japan) operating at 2.45 GHz was used for the sintering experiments. The sample was placed in a thermally insulated box, the inner surface of which was coated with SiC to act as a susceptor. The microwaves were generated by a magnetron with a 1.5-kW maximum rating at 2.45 GHz. The temperature was controlled by adjusting the voltage and current to induce a variable load on the magnetron. The sample temperature was measured by using an optical radiation thermometer. The density of the sintered samples was measured by the Archimedes method.

RESULTS AND DISCUSSION

Figure 1 and Figure 2 show the relationship between the amount of liquid-phase additive and the relative density of samples sintered at from 850 ℃ to 950 ℃ in case of premixed sample and reaction sintering process sample, respectively. The true density of the sample in each condition was calculated by summing the true density multiplied by the volume ratio of each raw material. The relative density was obtained by dividing experimental value by this calculated value. The density of sample which was made by reaction sintering with boric acid and the barium carbonate was plotted in solid line and one which was mixed barium borate was plotted in dash line respectively. In cases of premixed sample, the density of all samples became greatest in the vicinity of 925 ℃ that, were fluid phase generation temperature. As quantity of liquid phase component increased, the relative density became dense. However,

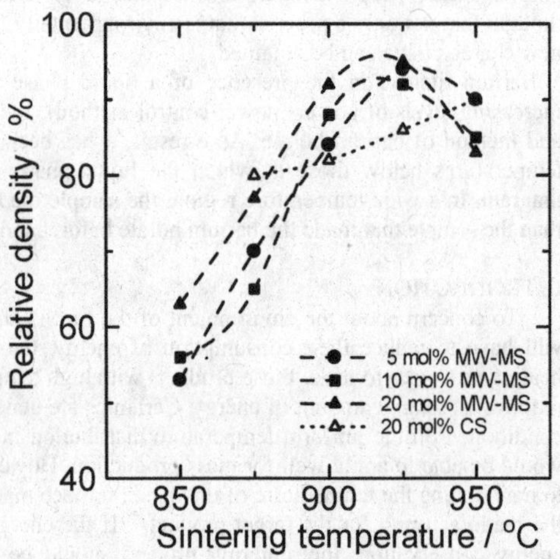

Fig. 1 Relative density of $BaTiO_3$ with liquid phase as a function of sintering temperature in case of premixed sample. MW-MS and CS mean microwave premixed sintering and conventional sintering, respectively.

a relative density of all samples decreased when the sintering temperature became 950 ℃ or more. On the other hand, the situation was different when reaction sintering was used. The sample was not densified at 900 ℃ when 5 mol% of liquid-phase material was added. The density increased to 90% at 950 ℃, which is above the temperature necessary for the generation of a liquid phase. When the amount of liquid-phase material was 10 mol%, a relative density of 80% was achieved at 850 ℃. The density increased with increasing temperature, reaching 97% at 900 ℃: it then remained almost constant at up to 950 ℃. When 20% of liquid phase was used, the densification behavior was different: after decreasing tendency in relative density between 850 ℃ and 875 ℃, the densification of the sample advanced at up to 900 ℃, as in the case of the sample containing 10 mol% of liquid phase. However, the density decreased when the sintering temperature exceeded 900 ℃. When we used a conventional electric furnace, a sample containing 20 mol% of BaB_2O_4 was not densified at 950 ℃.

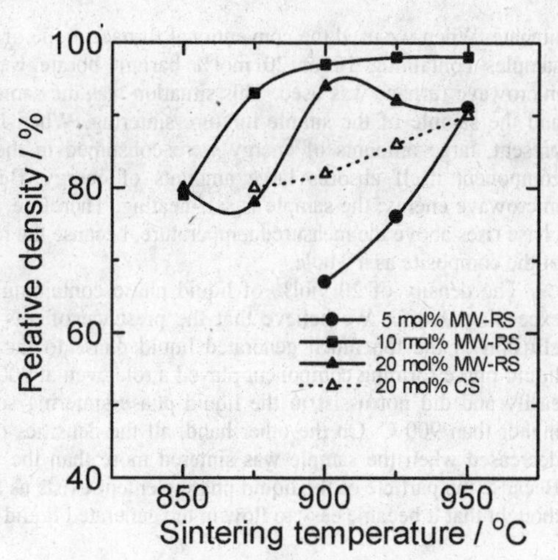

Fig. 2 Relative density of $BaTiO_3$ with liquid phase as a function of sintering temperature in case of reaction sintering process sample. MW-MS and CS mean microwave premixed sintering and conventional sintering, respectively.

Figure 3 shows the changes in the electric power consumed by the microwave furnace for samples of various compositions by using reaction sintering. The heating rate was 0.5 ℃/s and the sintering temperature was 900 ℃ for 1200 s. The least energy was consumed for the sample containing 5 mol% of liquid phase. Samples containing 10 or 20 mol% liquid phase showed almost equal power consumptions. However, electric power was consumed more efficiently at low temperatures by the sample containing 20 mol% liquid phase. Because the liquid phase absorbs the microwave easily, it is thought that the generation of the liquid phase became early.

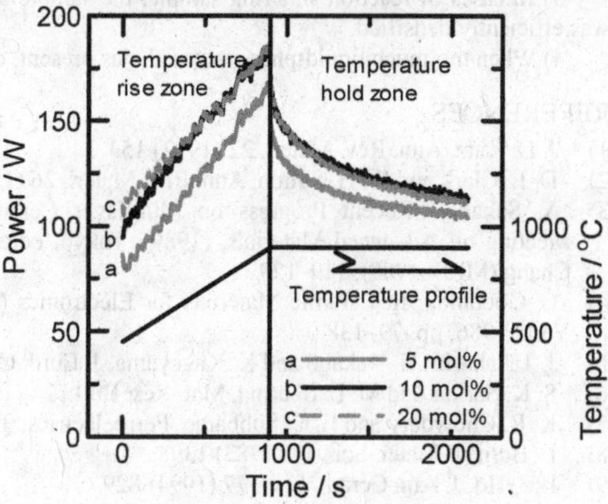

Fig. 3 Change in consumed power in the furnace with time.
Sintering condition:
Holding 900 °C for 1200 sec.
Rate of temperature increase 0.5 °C / sec

We consider the process for generating the liquid phase on the basis of the results that we obtained by using the microwave furnace. It is known that barium borate forms a liquid phase at 924 ℃ when mixed with barium titanate. When we used the conventional furnace, none of the samples was densified at 900 ℃, but the samples containing 10 or 20 mol% barium borate were densified at this temperature when the microwave furnace was used. This situation was the same also in the sample of the reaction sintering and the sample of the simple mixture sintering. When large amounts of liquid-phase material were present, large amounts of energy were consumed in the furnace. This shows that the liquid-phase component itself absorbs large amounts of energy. Because the liquid-phase component absorbs microwave energy, the sample is self-heating. Therefore, it is likely that the temperature of the liquid phase rises above the measured temperature, because the measured temperature records the temperature of the composite as a whole.

The density of 20 mol% of liquid phase containing sample decreased at sintering temperatures exceeding 900 ℃. We believe that the presence of lots of pores in the microstructure a result of effusion of the too much generated liquid phase to the outside of the sample. In other words, the liquid-phase forming component played a role even at 900 ℃. However, it flowed out from the sample easily and did not assist in the liquid phase sintering so the viscosity was lowered at temperatures higher than 900 ℃. On the other hand, all the densities of the sample of the simple mixture system decreased when the sample was sintered more than the temperature that the liquid phase generated. Because the particle of the liquid phase element exists as a big mass in the simple mixture system, it is thought that it became easy to flow in the generated liquid phase.

CONCLUSIONS

We obtained the following results when ceramics were sintered by liquid-phase microwave sintering.

1) The sample was densified at a temperature below the temperature where the liquid phase is normally generated when a sufficient amount of the liquid-phase material was present.

2) In cases of premixed sample, the density of all samples became greatest in the vicinity of 925 ℃

3) In cases of reaction sintering sample, the sample containing 10 mol% of liquid-phase material was efficiently densified.

4) When too much liquid-phase material was present, densification of the sample was obstructed.

REFERENCES

1) J. D. Katz, Ann. Rev. Mater., 22 (1992) 153.
2) D. E. Clark and W. H. Sutton, Ann. Rev. Mater., 26 (1996) 299.
3) Y. Sakabe, “Recent Progress on Multilayer Ceramic Capacitors”, Tenth MRS International Meeting on Advanced Materials, (1989), Tokyo, edited by M. Doyama, S. Somiya and R. P. H. Chang (MRS, 1989), 119–129.
4) G. Goodman, In Ceramic Materials for Electronics (Ed. R. C. Buchanan), Marcel Dekker, New York, 1986, pp. 79–138.
5) J. Takahashi, H. Nakano and K. Kageyama, J. Euro. Ceram. Soc., 26(2006), 2123
6) S. K. Sarkar and M. L. Sharma, Mat. Res. Bull., 24 (1989), 773
7) K. R. Chowdary and E. C. Subbarao, Ferroelectrics, 37 (1981) 689.
8) L. Burn, J. Mater. Sci., 17 (1982) 1398.
9) I. C. Ho, J. Am. Ceram. Soc., 77 (1994) 829.
10) D. Kolar, M. Trintelj, and L. Marsel, J. Phys. C1, 47 (1986), 447.
11) J. H. Lee, J. J. Kim, H. Wang and S. H. Cho, J. Mater. Res., 7 (2000) 1600.
12) Y. Goto and L. E. Cross, Yogyo-Kyokai-shi, 77 (1969) 355.

OPTIMIZATION OF MICROWAVE-ASSISTED RAPID DEBINDING OF CIM PARTS IN MULTIMODE APPLICATIONS

Roberto Rosa, Paolo Veronesi, Cristina Leonelli
Dipartimento di Ingegneria dei Materiali e dell'Ambiente, Università degli Studi di Modena e Reggio Emilia, via Vignolese 905, 41100, Modena, Italy

ABSTRACT

Microwave (MW) heating selectivity was exploited in this work for the rapid thermal debinding of parts obtained by Ceramic Injection Moulding (CIM). Since the organic binder preferentially absorbs microwaves with respect to ceramic powders, heat can be efficiently transferred to the green parts, despite their low thermal conductivity which renders conventional heating techniques less effective. However, one of the major drawbacks of microwave-assisted processes is the lack of reproducibility of the results, and of non-adequate experimental conditions and procedures which can lead to misleading conclusions on the effective yield of the process. A rational approach to overcome this problem, consisting in numerical simulation coupled to Design of Experiments (DoE) technique was used. In this way it has been possible to optimize the MW-assisted thermal debinding of ring-shaped CIM parts in multi-mode applicators operating at 2.45 GHz, reducing processing times from the original 80-140 hours (depending on the shape and dimensions of the samples) by conventional heating to 6 hours by dielectric heating.

INTRODUCTION

During forming of technical ceramic green parts by Ceramic Injection Moulding (CIM), it is necessary to use an organic binder, which can be of different nature and added in different quantities (up to 40 wt%). The binder must be removed prior to sintering, in order to avoid defects or breakage of the final products. The industrial debinding is typically conducted with chemical methods (exploiting for example solvents[1] or supercritical CO_2[2]), thermal methods (exploiting conventional[3] or microwave[4,5] heating) or with a combination of both[6].

Thermal debinding requires an appropriate debinding temperature (generally lower than 600°C), depending on the organic binder nature and a slow and accurately controlled heating. In fact a too rapid heating could induce overpressures in the green parts, inducing distortions, cracking and porosity formation[7]. When heat is generated in the load by dielectric heating, these requirements result particularly demanding, but proper microwave-matter interaction conditions could help increasing heating homogeneity in the green parts. As a matter of fact, the possibility of controlling accurately the electromagnetic field distribution in the load and consequently the heat generation on the basis of dielectric properties of the load makes microwave an attractive technology to perform thermal debinding, also in combination with other heating methods (resistance heating, infrared, hot air,...).

Power density distribution generated in the material, as a consequence of the MW-matter interaction, is given by Equation 1:

$$P_d(x,y,z) = \omega\, \varepsilon_0 \varepsilon''_{eff} E_{rms}^2 + \omega\, \mu_0 \mu''_{eff} H_{rms} \quad (1)$$

where:
P_d = power density in the material (W/m^3), at the position (x,y,z),
$\omega = 2\pi f$ (Hz), f = frequency of the incident microwaves,
ε''_{eff} = effective loss factor, including conductivity losses,
μ''_{eff} = imaginary part of the effective magnetic permeability,
E_{rms}= local (x,y,z) electric field intensity (V/m),
H_{rms}= local (x,y,z) magnetic field intensity (A/m).

Therefore, as shown in *Equation 1*, depending on the dielectric properties of the organic binder and of the ceramic powders, as well as on the electromagnetic field distribution inside the material, heat can be preferentially generated in the load. In particular, since currently used organic binders present a loss tangent higher than most oxide-based ceramic materials at temperatures lower than 600℃ [8], heat selectivity can be successfully achieved, and the binder is thus heated volumetrically in the whole CIM part. This selective heating is expected to lead to a faster thermal debinding, which has no longer to rely on the slow heat transfer by conduction and convection. However, when processing simultaneously multiple CIM parts, the use of medium-size multi-mode microwave applicators is almost mandatory, which means that there is a strong possibility of having not homogenous electromagnetic field distributions in the load, and, according to *Equation 1*, uneven heat generation in the load. For this reason, when dealing with a microwave multi-mode applicator, one of the major drawbacks regards the poor reproducibility of the treatment, if not properly controlled. In this framework, numerical simulation is a useful tool to investigate the local electromagnetic field distribution inside the load, allowing to determine the most favourable microwave-matter interactions[9]. Modeling large and complex systems however, requires enormous computational resources, and consequently, time. Thus in the present study, in order to reduce the number of computational simulations, Design of Experiments (DoE) techniques were applied, performing the experiments in a "virtual" environment, namely the simulated multi-mode applicator. In our case, being the experiments indicated by DoE firstly performed by numerical simulations, we refer to our approach as Virtual Design of Experiments.

EXPERIMENTAL

Ring-shaped CIM green parts made of Al_2O_3 containing up to 35 wt% of organic binder, were used for the microwave assisted thermal debinding process. To obtain input data for numerical simulations, room temperature dielectric properties of the green parts were measured in the 1-3 GHz frequency range using an Agilent 8753D Vector Network analyzer connected to the Agilent 85070E dielectric probe kit. Smooth disc-shaped samples were prepared in order to optimise the contact between the probe and the material during the dielectric properties measurements. In order to reduce the number of numerical simulations, needed to reach the most possible optimized configuration, in terms of process speed, heating homogeneity and energy efficiency, the software Design Expert v.6 was used. The independent variables of the studied process, inserted in the software were:

- number of samples positioned in the MW cavity (from 4 to 20),
- reciprocal distance between each sample (from 2 to 12 mm),
- refractory materials (microwave absorbing SiC element or microwave transparent Al_2O_3 fibre board).

Once the software returned the experimental plan needed to investigate variables interactions, an overall of 18 virtual experiments were performed in order to determine the system responses.

The software Concerto 4.0 (Vector Fields, U.K.) was used to firstly perform experiments trough numerical simulation of the electromagnetic field distributions, in a cavity simulating the CEM-MAS 7000 (CEM, U.S.A.) furnace. The model used to simulate the CEM-MAS 7000 cavity is shown in Figure 1. Load materials with their equivalent dielectric properties measured at room temperature and at 2.45 GHz are reported in Table I. Output data obtained from numerical simulations were the volumetric Specific Absorption Rate (SAR, in W/kg) distribution, form which mean value, variance and gradient were calculated in each green part and in the whole load. The total dissipated power (W) was also extracted from the model results and used as a measure of the overall energy efficiency of the process. All the output data from 18 numerical simulations were then re-inserted in the DoE software to obtain the optimised model with the highest desirability. Experimental validation of the optimised debinding conditions of CIM green samples were performed in a CEM-MAS 7000 microwave furnace.

The non-homogeneous electromagnetic field distribution in the MW cavity does not assure that all CIM parts are subjected to similar heating treatments and for this reason preliminary heating tests up to 250°C were performed measuring temper ature with four Neoptix Reflex optical fibres placed in contact with four different samples each run. Preliminary tests allowed to conclude that no relevant temperature differences could be detected for different parts, in particularly when the MW co-absorber (C-shaped SiC element) was used. Optimised treatments temperature was measured by means of a single Neoptix Reflex optical fibre in the temperature range 20-250°C and with a Mikron sapphire fibre, to assure the achievement and the maintenance of the debinding temperature of 600°C.

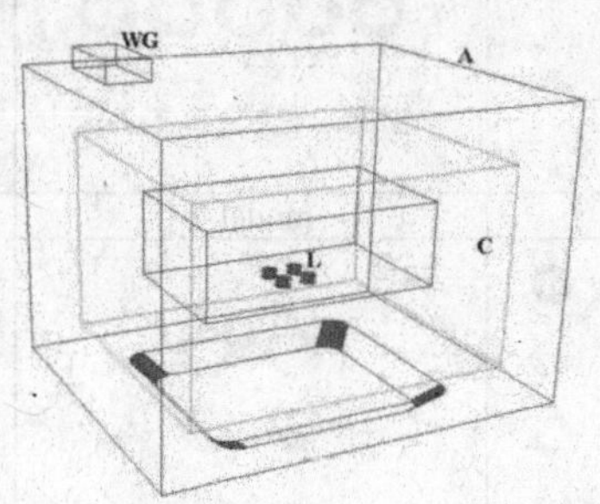

Figure 1 – Model of the applicator (A) with the microwave feeding waveguide (WG), refractory chamber (C) and load position (L). Applicator dimensions are approximately 400 x 330 x 280 mm.

Table I – Model materials with lettering according to Figure 1 and equivalent dielectric properties at 2.45 GHz; (m)= measured

Part	Material	Dielectric properties[8, 10, 11]
Applicator walls (A)	Perfect electric conductor	
Waveguide (WG)	Perfect electric conductor	
Refractory chamber (C)	Alumina fibre	9.5-j0.000285
Refractory plates	Alumina or SiC	8.9-j0.009 or 30-j11
CIM alumina (green parts) (L)	Al_2O_3 + 25 wt% binder	12-j0.2 (m)

RESULTS AND DISCUSSION

In Table II are reported the independent variable values for the 18 models selected by DoE software. On the basis of DoE experimental plan, 18 numerical simulations were performed.

Table II. DoE experimental plan in terms of number of parts, reciprocal distance between each part and refractory board material (S = SiC, A = alumina)

Model #	1	2	3	4	5	6	7	8	9	10	11	12	13	14	15	16	17	18
number of parts	4	20	4	20	4	20	12	12	12	4	20	4	20	4	20	12	12	12
reciprocal distance (mm)	2	2	12	12	7	7	2	12	7	2	2	12	12	7	7	2	12	7
refractory board material	S	S	S	S	S	S	S	S	S	A	A	A	A	A	A	A	A	A

Figure 2 and 3 show results in terms of SAR distribution in the middle layer of xy and xz planes for samples treated in the presence of C-shaped silicon carbide MW co-absorber and of an Al_2O_3 refractory plate respectively.

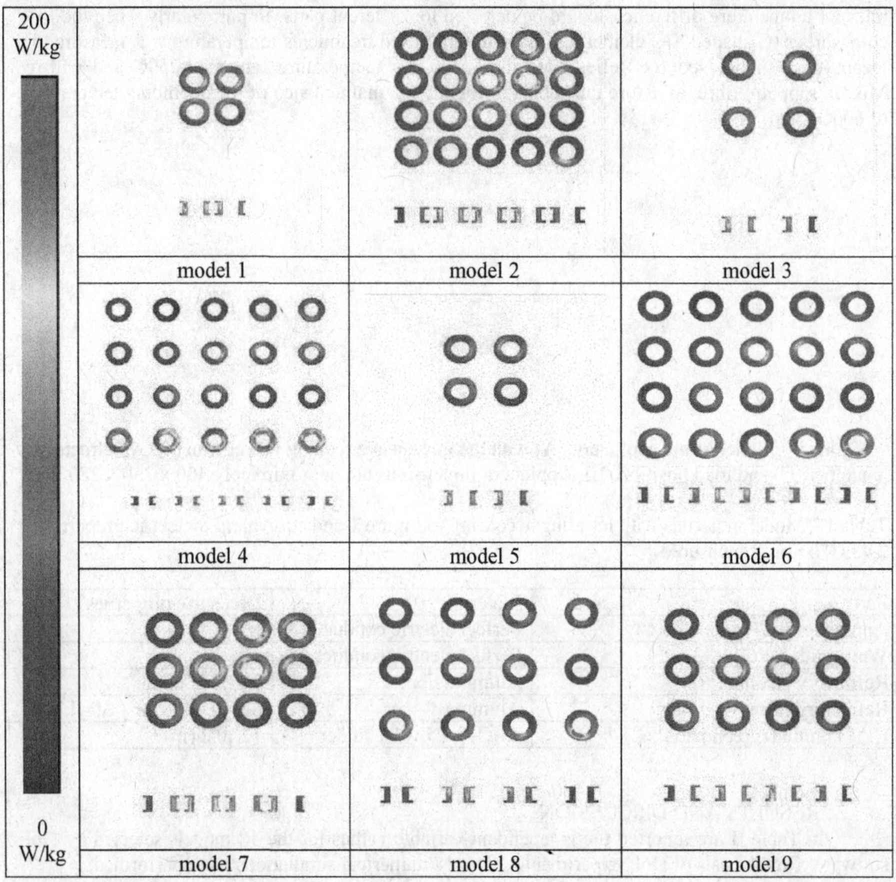

200 W/kg

model 1 | model 2 | model 3

model 4 | model 5 | model 6

0 W/kg

model 7 | model 8 | model 9

Figure 2. SAR distribution in the middle layer of xy (each figure top) and xz (each figure bottom) planes obtained from numerical simulations in the presence of C-shaped SiC element, as MW co-adsorber.

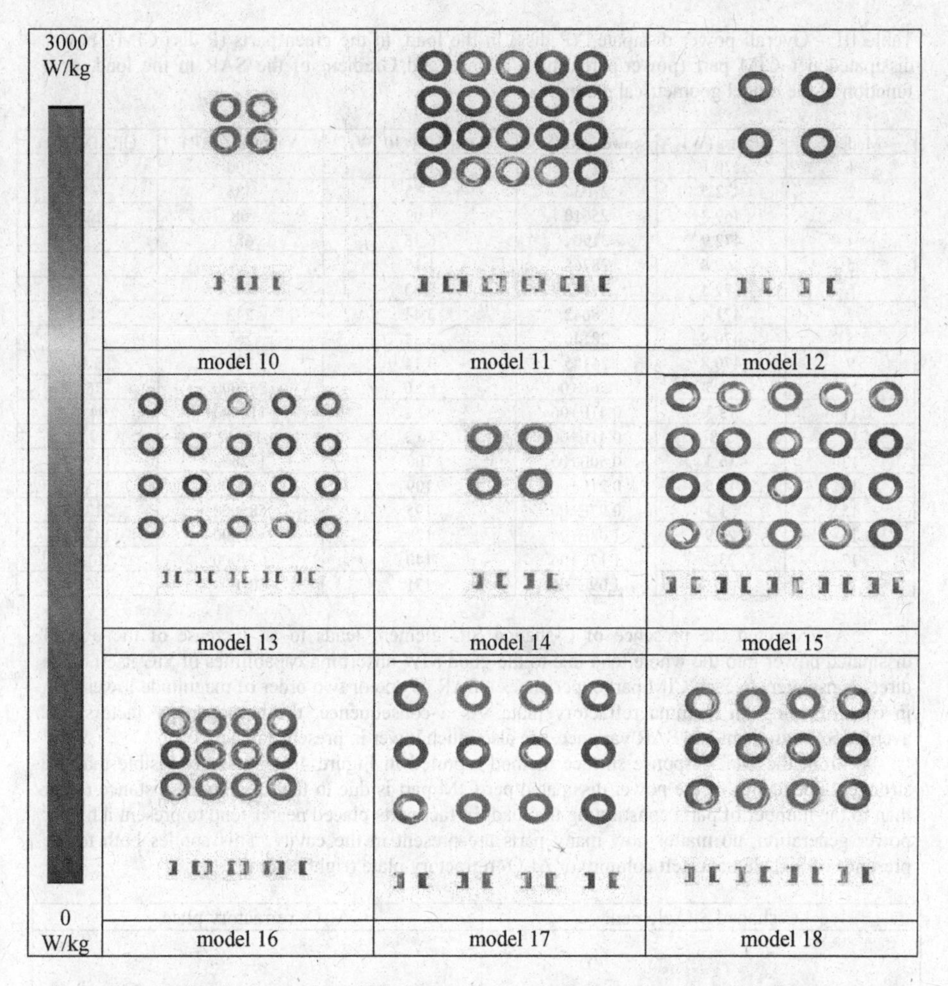

Figure 3. SAR distribution in the middle layer of xy (each figure top) and xz (each figure bottom) planes obtained from numerical simulations in the presence of Al_2O_3 refractory support.

As it is clearly visible by two different ranges for SAR distribution in Figures 2 and 3, C-shaped SiC element surrounding the samples into the MW cavity, strongly attenuates the electromagnetic energy directed on the CIM parts. Silicon carbide presents a high MW absorbing behaviour, as well as a high thermal conductivity. For this reason SiC is expected to homogenise the temperature distribution also due to its capability of rapidly conducting heat towards the colder regions.

From the SAR 3D distribution, the quantities shown in Table III have been calculated.

Table III – Overall power dissipated (P diss) in the load, in the green parts (P diss CIM), power dissipated per CIM part (power/part) and Variance and Gradient of the SAR in the load, as a function of the model geometrical parameters.

Model #	P diss (W)	power/part (W)	Pdiss CIM (10^5 W)	Variance (SAR)	GRAD(SAR)
1	470.1	31650	1.27	100	1.54
2	472.5	27634	5.53	385	6.58
3	469.8	25610	1.02	68	1.24
4	472.9	31904	6.38	681	7.58
5	469.8	28768	1.15	83	1.39
6	472.4	.29172	5.83	472	6.68
7	471.4	28642	3.44	243	4.1
8	470.9	28140	3.38	267	4
9	470.8	26185	3.14	203	3.59
10	10.5	1.56E+06	62.6	265000	75.7
11	13.3	0.41E+06	82.2	116000	94.91
12	2.8	0.41E+06	16.3	18262	19.28
13	16.3	0.50E+06	100	172000	115.77
14	18.5	0.21E+06	109	829000	131.12
15	31.1	0.97E+06	195	585000	217.38
16	20.9	1.03E+06	123	407000	143.12
17	23.2	1.17E+06	140	674000	161.62
18	21	1.09E+06	131	507000	146.28

As expected the presence of C-shaped SiC element leads to an increase of the overall dissipated power into the whole load due to the good MW absorbing capabilities of SiC itself. As a direct consequence, each CIM part experiences a SAR of one or two order of magnitude lower than in case of using an alumina refractory plate. As a consequence, the homogeneity factors, i.e. average SAR gradient and SAR variance, are also much lower in presence of SiC.

From the DoE response surface method reported in Figure 4, it is clearly visible that the stronger dependence of the power dissipated per CIM part is due to their reciprocal distance rather than to the number of parts constituting the load. In fact parts placed nearer tend to present a higher power generation, no matter how many parts are present in the cavity. This applies both to the presence of SiC element (left column) or Al_2O_3 refractory plate (right column).

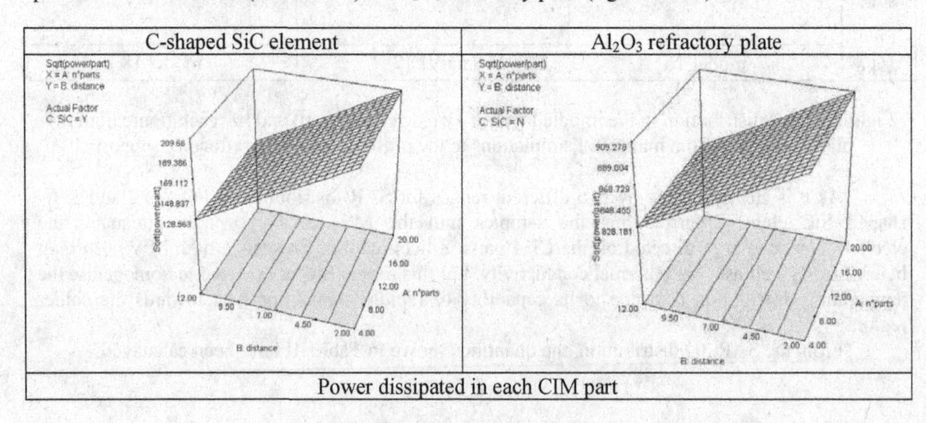

Figure 4. Power dissipated in each CIM part.

A possible explanation lies in the electromagnetic field distribution inside the modelled furnace, which tends to be concentrated in the central regions of the microwave applicator. Thus, in case of parts placed at close distance, they tend to be positioned near the applicator central part, and hence being subjected to higher power generation. Through Virtual Design of Experiment approach, it has been possible to determine the optimised debinding treatment, in terms of heating homogeneity and energy efficiency; it consisted in a load configuration made of 20 CIM parts, spaced 2 mm and surrounded by the C-shaped SiC element. This configuration corresponds to model 2 in Figure 2. Desirability of this configuration as obtained by DoE is 50.4%.

Figure 5 shows the expected 3D SAR distribution in case of optimum treatment conditions.

Figure 5. Expected 3D SAR distribution in case of optimum load configuration made of 20 CIM parts, spaced 2mm and surrounded by the C-shaped SiC element.

This optimised load disposition was then experimentally validated in a multi-mode CEM MAS 7000 microwave furnace. The resulting debinding time using microwave was found to be 6 hours, in case of 5 mm thick rings. Debinding times in conventional industrial processes for such samples are in the 80-140 hours range, depending on the shape and dimensions of the samples.

In the attempt to further reduce the debinding time in the MW assisted approach, blisters and distortions of samples appeared. It is noticeable that it was impossible to obtain non cracked samples when applying conventional heating cycles similar to the microwave-assisted ones.

Figure 6 shows the green parts after microwave assisted debinding in the optimised conditions.

Figure 6. Ring-shaped alumina samples after 6 hours debinding in microwave multi-mode applicator. On the top left two examples of green parts before microwave assisted debinding.

CONCLUSIONS

DoE technique coupled to numerical simulation of electromagnetic field distribution in a microwave multi-mode applicator allowed to drastically reduce time required to reach the optimization of MW-assisted debinding process of CIM green parts. In fact, the so-called virtual Design of Experiments approach allows to find the best load configuration in terms of number of CIM parts, reciprocal distances between each sample and supporting/surrounding material to be used. In particular, in case of ring-shaped alumina samples, the optimum load arrangement consisted of the maximum number of parts, positioned at the minimum distance from each other and surrounded by a C-shaped silicon carbide MW co-absorber.

Computational results were subsequently confirmed by experimental validation. Microwave assisted debinding of CIM parts in a multi-mode CEM MAS 7000 applicator led to a reduction from 140 to 6 hours with respect to conventional industrial processes.

REFERENCES

[1]V. A. Krauss, A. A. M. Oliveira, A. N. Klein, H. A. Al-Qureshi and M. C. Fredel, A Model for PEG Removal from Alumina Injection Moulded Parts by Solvent Debinding, *J. Mater. Process. Technol.* **182**, 268-273 (2007).

[2]Y. Wu, W.-J. Si and H.-Z. Miao, Kinetics for Supercritical CO_2 Debinding of Injection Molded ZrO_2, *Key Eng. Mater.* **368-372**, 736-739 (2008).

[3]M. Descamps, T. Duhoo, F. Monchau, J. Lu, P. Hardouin, J. C. Hornez and A. Leriche, Manufacture of Macroporous β-Tricalcium Phosphate Bioceramics, *J. Eur. Ceram. Soc.* **28**, 149-157 (2008).

[4]C. Leonelli, G. C. Pellacani, C. Siligardi and P. Veronesi, Microwave Assisted Burn-Out of Organic Compounds in Ceramic Systems, *Key Eng. Mater.* **264-268**, 739-742 (2004).

[5]Z. Xie, Y. Huang, J. Wu and L. Zheng, Microwave Debinding of a Ceramic Injection Molded Body, *J. Mater. Sci. Letters* **14 (11)**, 794-795 (1995).

[6]P. Thomas-Vielma, A. Cervera, B. Lavenfeld and A. Varez, Production of Alumina Parts by Powder Injection Molding with a Binder System Based on High Density Polyethylene, *J. Eur. Ceram. Soc.* **28**, 763-771 (2008).

[7]W. J. Tseng and C.-K. Hsu, Cracking Defect and Porosity Evolution During Thermal Debinding in Ceramic Injection Moldings, *Ceram. Int.* **25**, 461-466 (1999).

[8]A. C. Metaxas and R.J. Meredith, *Industrial Microwave Heating*, Peter Peregrinus, London (1983).

[9]P. Veronesi, C. Leonelli, M. R. Rivasi and G. C. Pellacani, The Electromagnetic Field Modeling as a Tool in the Microwave Heating Feasibility Studies, *Mater. Res. Innov.* **8 (1)**, 9-12 (2004).

[10]R. F. Schiffmann, Principles of Industrial Microwave and RF Heating, in *Microwaves: Theory and Applications in Materials Processing IV*, *Ceramics Transactions* **80**, 41-60 (1997).

[11]M. Sato, Insulation Blankets for Microwave Sintering of Traditional Ceramics, in *Microwaves: Theory and Applications in Materials Processing V*, *Ceramics Transactions* **111**, 277-285 (2001).

CHARACTERIZATION OF A POTENTIAL SUPERPLASTIC ZIRCONIA–SPINEL NANOCOMPOSITE PROCESSED BY SPARK PLASMA SINTERING

Mahmood Shirooyeh[1], Sohana Tanju[2], Javier E. Garay[2], Terence G. Langdon[1,3]

[1] Departments of Aerospace & Mechanical Engineering and Materials Science, University of Southern California, Los Angeles, CA 90089-1453, U.S.A.
[2] Department of Mechanical Engineering, University of California, Riverside, CA 92521, U.S.A.
[3] Materials Research Group, School of Engineering Sciences, University of Southampton, Southampton SO17 1BJ, U.K.

ABSTRACT

Ceramics based on the Y_2O_3-stabilized tetragonal zirconia system are well-known for exhibiting superplastic properties in tensile testing at elevated temperatures. It is also known that in order to attain High Strain Rate Superplasticity (HSRS) in zirconia ceramics it is necessary both to suppress grain growth during sintering and to enhance cation diffusion. The present investigation was initiated to evaluate the potential for producing a zirconia-spinel nanocomposite suitable for achieving superplasticity using a processing method based on Spark Plasma Sintering (SPS) where this is a very rapid electric current-activated sintering technique having a heating rate of 300℃/min. The experiments show that it is possible to achieve almost fully-dense tetragonal ZrO_2 nanocomposites dispersed with 30 vol% $MgAl_2O_4$ at the relatively low processing temperature of 1523 K by using a combination of high-energy ball-milling and SPS. Microstructural observations were made after processing and they revealed an average grain size of the order of ~100 nm. It is shown using microhardness measurements that the disks produced using SPS are essentially homogeneous across their diameters thereby suggesting these nanocomposite materials probably have a potential for exhibiting excellent superplastic behavior.

INTRODUCTION

Superplasticity refers to the ability of some materials to pull out, in a generally isotropic manner, to exceptionally high elongations when testing in tension. Typically, superplastic flow is considered as tensile elongations equal to or greater than 500%.

Although superplastic behavior was reported in many metallic alloys over a period of many years, it was generally considered that ceramic materials are inherently brittle and therefore they appeared to be incapable of exhibiting high tensile ductilities. This belief changed on July 5, 1985, when Japanese newspapers (*Asahi Shimbun*, *Nippon Keizai Shimbun* and others) reported the remarkable tensile ductilities achieved by Wakai and co-workers in experiments on an yttria-stabilized tetragonal polycrystalline zirconia (generally designated as 3Y-TZP where the number preceding Y denotes the mole percentage of yttria). Specifically, this early work showed an elongation of >120% which, although not large by comparison with superplastic metals, was nevertheless unusually high for a ceramic material. A report of this work appeared in 1986 in the scientific literature[1] and subsequently later reports documented elongations as high as ~800% in 3Y-TZP[2] and ~1038% in 2.5Y-TZP containing 5 wt % SiO_2.[3] A later report described a tensile elongation of ~1050% in a composite of zirconia, alumina and spinel when testing at a temperature of 1923 K at a strain rate of 1 s^{-1} which is exceptionally rapid for this type of flow.[4] Since High Strain Rate Superplasticity (HSRS) is defined formally as superplastic flow occurring at strain rates at and above 10^{-2} s^{-1}, it is clear that this latter result falls within the HSRS regime.[5] An early review described the occurrence of superplasticity in ceramics[6] and a later review summarized the superplastic behavior of a range of materials including metals, ceramics and intermetallic compounds.[7]

Superplastic flow in metals is now well documented in the scientific literature. Flow occurs by the mechanism of grain boundary sliding in which individual grains in the polycrystalline matrix move over each other with the deformation occurring at, or immediately adjacent to, the boundary plane.[8] However, a polycrystalline matrix cannot deform solely by grain boundary sliding without opening up large voids and in practice some concomitant accommodation process must occur within the grains. The basic model for superplastic flow in metals involves sliding accommodated by some limited slip within the adjacent grains,[9] where this is consistent with direct experimental evidence for the accommodation process.[10] The situation in ceramics may be different because an analysis suggests the anticipated yield stresses for dislocation slip are significantly higher than the flow stresses measured experimentally in superplastic experiments on 3Y-TZP.[11] This has led to an alternative proposal for these ceramic materials in which superplasticity occurs through the occurrence of Coble diffusion creep controlled by movement of the Zr^{4+} ions.[11] The latter proposal is consistent with a detailed analysis of the flow behavior.[12,13]

The present research was initiated to evaluate the potential for using a combination of ball-milling and Spark Plasma Sintering (SPS) to produce a fully-dense tetragonal zirconia nanocomposite containing 30 vol% $MgAl_2O_4$ that may be suitable for exhibiting superplastic characteristics. Earlier investigations showed the possibility of using SPS for the production of superplastic ceramics and nanocomposites[14-17] and it is well known that excellent superplastic properties are often achieved in yttria-stabilized zirconia through the use of doping.[18,19]

EXPERIMENTAL MATERIAL AND PROCEDURES

The experiments were conducted using two starting materials: a high-purity 3 mol.% Y_2O_3-stabilized tetragonal ZrO_2 nanopowder (TZ-3Y, Tosoh Co. Ltd., Tokyo, Japan) with a ~40 nm particle size and a specific surface area of 16 $m^2.g^{-1}$ and a high-purity magnesium aluminate ($MgAl_2O_4$) spinel nanopowder (American Elements, CA, USA) with an average particle size of ~30 nm. Milling was achieved using an attritor, a high-energy ball mill used in mechanical alloying. The powder was milled for 2 h with ethanol to prepare a composite powder. In order to prevent any extraneous contamination, the grinding media and the container of the attritor were made from zirconia. Following milling, the mixture was dried, crushed and sieved in order to yield a homogeneous particle size.

The powder mixtures were placed into a graphite die having an inner diameter of 19 mm and then subjected to different loads of 20 kN or 30 kN corresponding to nominal pressures of 70 and 106 MPa, respectively. During application of the pressure, the specimens were heated from room temperature with a heating rate of 300°C/min to different sintering temperatures of either 1473 or 1523 K using a Spark Plasma Sintering (SPS) system under vacuum. A schematic illustration of the SPS facility is shown in Fig. 1. The total time for processing was 700 s to produce a ceramic disc with a thickness of 1-2 mm consisting of Y-TZP-30 vol % $MgAl_2O_4$. The discs were then cooled to room temperature. Throughout the processing operation, the temperature was controlled and monitored using a thermocouple located through a hole at the external wall of the graphite die.

Following processing by SPS, the relative densities of the specimens were measured using the Archimedes method with the theoretical density of the composite calculated as 5.31×10^{3} $kg.m^{-3}$ based on the conventional rule of mixtures.

In order to obtain information on the homogeneity of the microstructures within each disc, microhardness measurements were conducted using a Vickers microhardness tester with an indentation load of 1000 g and a dwell time of 15 s. The hardness measurements were taken on the cross-sectional areas of the specimens. For microstructural observation, the specimens were mounted, mechanically polished and relief etched using colloidal silica. A thermal field emission scanning electron microscope (SEM) with a resolution better than 2 nm (JEOL JSM-7001F) was used for microstructural characterization. The microscope was equipped with an Energy Disperse Spectrometer (EDS).

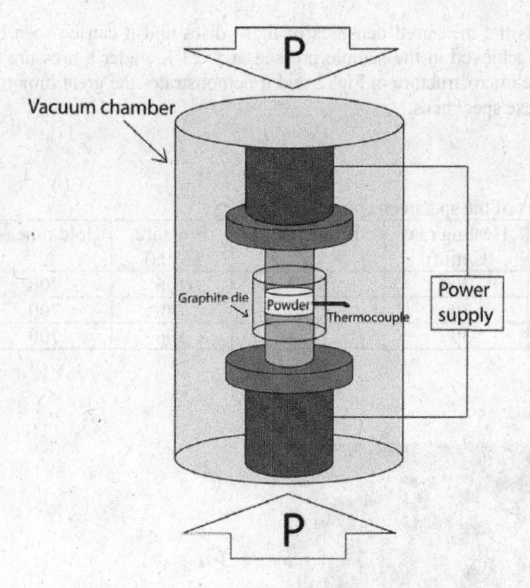

Figure 1. Schematic illustration of the SPS system

EXPERIMENTAL RESULTS AND DISCUSSION

For the spark plasma sintering, different temperatures and pressures were used in order to evaluate the effect of temperature and pressure on the density and homogeneity of the microstructures. The range of temperatures and pressures were determined based on previous experience with zirconia ceramics.[20-22] It was shown an earlier investigation that pressure has the greatest influence on the final density of the nanoscale zirconia ceramics.[20]

Figure 2 shows a representative SEM micrograph and EDS spectrum of the microstructure of the sample processed by SPS at 1523 K with a pressure of 106 MPa. An average grain size of ~100 nm was estimated for this specimen based on the fracture surface in high-resolution SEM micrographs. This is smaller than the reported grain sizes of ~290 nm and ~420 nm for zirconia and spinel phases, respectively, obtained via a pressureless sintering process with a higher sintering temperature.[23]

In order to attain high strain rate superplasticity in ceramics, it is known that a nanoscale grain size is critical. In the present study, it is important to note that very limited grain growth was observed. In Fig. 2, the microstructural observations show a dense zirconia matrix with a well-dispersed spinel phase across the specimen. The dispersion of the second phase is also of considerable importance in obtaining favorable microstructures. In Fig. 2 the light-colored phase corresponds to the zirconia phase and the darker phase was identified as the spinel phase. The EDS analysis, also shown in Fig. 2, confirms the presence of the two main phases of zirconia and spinel with no evidence for any extraneous impurities. It should be noted that the carbon peak in the EDS spectrum comes from the epoxy in which the specimen was embedded.

Table 1 records the measured densities of these discs and it can be seen that a relative final density of >95% was achieved in the sample pressed at 1523 K under a pressure of 106 MPa. This also corresponds to the microstructure in Fig. 2 and it demonstrates the great importance of pressure in the densification of these specimens.

Table 1. Characteristics of the specimens processed by SPS.

Material	Heating rate (°C/min)	Temperature (K)	Pressure (MPa)	Hold time (s)	Theoretical Density %
ZrO_2-30 vol% spinel	300	1473	106	700	>94
ZrO_2-30 vol% spinel	300	1523	70	700	>91
ZrO_2-30 vol% spinel	300	1523	106	700	>95

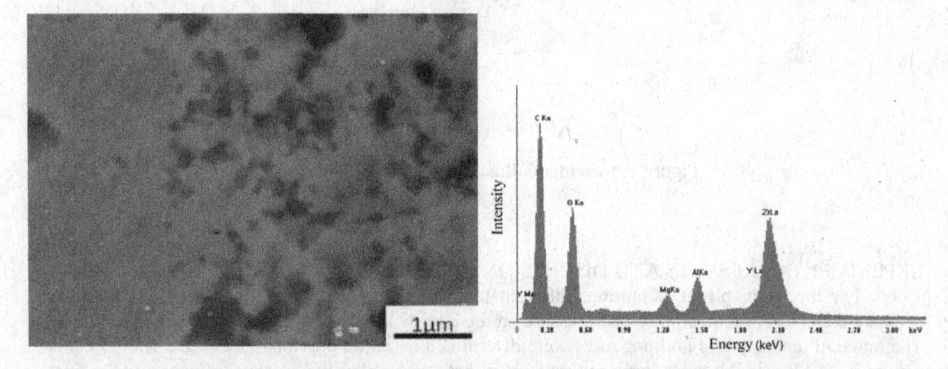

Figure 2. An SEM micrograph and an EDS spectrum of the Y-TZP-30 vol% spinel composite after processing by SPS at 1523 K under a pressure of 106 MPa.

Finally, Fig. 3 shows the values of the Vickers microhardness, Hv, for the two samples processed by SPS at 1523 K with the individual values of Hv plotted against the distance from the centers of the discs. These results are very encouraging because both sets of data confirm that the microstructure is very homogenous because the hardness remains practically constant across both diameters. Although the results from the two discs shown in Fig. 3 are similar, there is some evidence that a higher pressure during SPS may result in specimens having higher values of the Vickers microhardness.

The overall results from these experiments are encouraging. It is shown that a combination of high-energy ball milling and SPS is capable of producing excellent samples of a nanoscale zirconia-30 vol% spinel composite. The measured high densities and excellent homogeneities suggest these materials would be favorable for measuring superplastic properties at elevated temperatures.

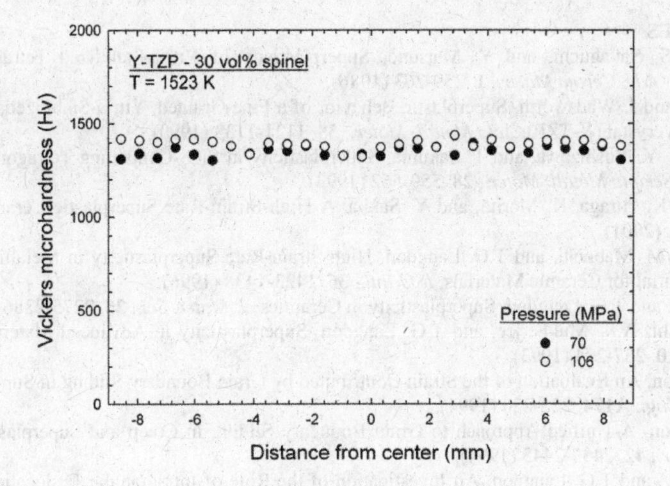

Figure 3. Variation of the average Vickers microhardness, Hv, with distance from the center of the samples processed by SPS.

SUMMARY AND CONCLUSIONS

Experiments were undertaken to evaluate the potential for using ball milling and spark plasma sintering to produce a nanoscale composite, in the zirconia-spinel system having properties favorable for exhibiting superplastic characteristics. The following conclusions follow directly from this investigation:

(1) A very rapid electric current-activated sintering technique, or spark plasma sintering, with a heating rate of 300℃/min and pressures up to 105 MPa was employed to produce a nanoscale ZrO_2- 30 vol% spinel ceramic composite at a relatively low temperature of 1473 K in a total time of 700 s.

(2) The results show it is possible to achieve very high density composites, with relative densities up to >95%, and with true nanoscale grain sizes in the range of ~100 nm.

(3) Detailed microstructural observations reveal a well-dispersed spinel phase across the specimens and arrays of small grains that are appropriate for achieving superplastic flow.

(4) By taking detailed values of the Vickers microhardness across sintered discs, it is shown that the materials exhibit excellent microstructural homogeneity.

ACKNOWLEDGEMENTS

We would like to sincerely thank Dr. R. P. Dillon (University of California, Irvine) for his assistance with the sample preparation. The high-energy ball milling facilities were kindly made available by Professor Martha L. Mecartney of the Department of Chemical Engineering and Materials Science, University of California, Irvine.

REFERENCES

[1]F. Wakai, S. Sakaguchi, and Y. Matsuno, Superplasticity of Yttria-Stabilized Tetragonal ZrO_2 Polycrystals, *Adv. Ceram Mater.*, **1**, 259-263 (1986).
[2]T.G. Nieh, and J. Wadsworth, Superplastic Behavior of a Fine-Grained, Yttria-Stabilized, Tetragonal Zirconia Polycrystal (Y-TZP), *Acta Metall. Mater.*, **38**, 1121-1133 (1990).
[3]K. Kajihara, Y. Yoshizawa, and T. Sakuma, Superplasticity in SiO_2-Containing Tetragonal Zirconia Polycrystal, *Scripta Metall. Mater.*, **28**, 559-562 (1993).
[4]B.-N. Kim, K. Hiraga, K. Morita, and Y. Sakka, A High-Strain-Rate Superplastic Ceramic, *Nature* **413**, 288-291 (2001).
[5]K. Higashi, M. Mabuchi, and T.G. Langdon, High-Strain-Rate Superplasticity in Metallic Materials and the Potential for Ceramic Materials, *ISIJ Intl.*, **36**, 1423-1438 (1996).
[6]Y. Maehara, and T.G. Langdon, Superplasticity in Ceramics, *J. Mater. Sci.*, **25**, 2275-2286 (1990).
[7]A.H. Chokshi, A.K. Mukherjee, and T.G. Langdon, Superplasticity in Advanced Materials, *Mater. Sci. Eng.*, **R10**, 237-274 (1993).
[8]T.G. Langdon, An Evaluation of the Strain Contributed by Grain Boundary Sliding in Superplasticity, *Mater. Sci. Eng.*, **A174**, 225-230 (1994).
[9]T.G. Langdon, A Unified Approach to Grain Boundary Sliding in Creep and Superplasticity, *Acta Metall. Mater.*, **42**, 2437-2443 (1994).
[10]R.Z. Valiev, and T.G. Langdon, An Investigation of the Role of Intragranular Dislocation Strain in the Superplastic Pb-62% Sn Eutectic Alloy, *Acta Metall. Mater.*, **41**, 949-954 (1993).
[11]M.Z. Berbon, and T.G. Langdon, An Examination of the Flow Process in Superplastic Yttria-Stabilized Tetragonal Zirconia, *Acta. Mater.*, **47**, 2485-2495 (1999).
[12]N. Balasubramanian, and T.G. Langdon, Comment on the Role of Intragranular Dislocations in Superplastic Yttria-Stabilized Zirconia, *Scripta Mater.*, **48**, 599-604 (2003).
[13]N. Balasubramanian, and T.G. Langdon, Flow Processes in Superplastic Yttria-Stabilized Zirconia: A Deformation Limit Diagram, *Mater. Sci. Eng.*, **A409**, 46-51 (2005).
[14]G.-D. Zhan, J. Kuntz, J. Wan, J. Garay, and A.K. Mukherjee, Spark-Plasma-Sintered $BaTiO_3/Al_2O_3$ Nanocomposites, *Mater. Sci. Eng.*, **A356**, 443-446 (2003).
[15]G.-D. Zhan, J.E. Garay, and A.K. Mukherjee, Ultralow-Temperature Superplasticity in Nanoceramic Composites, *Nano Lett.*, **5**, 2593-2697 (2005).
[16]D.M. Hulbert, D. Jiang, J.D. Kuntz, Y. Kodera, and A.K. Mukherjee, A Low-Temperature High-Strain-Rate Formable Nanocrystalline Superplastic Ceramic, *Scripta Mater.*, **56** (2007) 1103-1106.
[17]D. Jiang, D.M. Hilbert, J.D. Kuntz, U. Anselmi-Tamburini, and A.K. Mukherjee, Spark Plasma Sintering: A High Strain Rate Low Temperature Forming Tool for Ceramics, *Mater. Sci. Eng.*, **A463** (2007) 89-93.
[18]Y. Sakka, T. Ishii, T.S. Suzuki, K. Morita, and K. Hiraga, Fabrication of High-Strain Rate Superplastic Yttria-Doped Zirconia Polycrystals by Adding Manganese and Aluminum Oxides, *J. Eur. Ceram. Soc.*, **24**, 449-453 (2004).
[19]Y. Natanzon, M. Boniecki, and Z. Łodziana, Influence of Elastic Properties on Superplasticity in Doped Yttria-Stabilized Zirconia, *J. Phys. Chem. Solids*, **70**, 15-19 (2009).
[20]U. Anselmi-Tamburini, J.E. Garay, Z.A. Munir, A. Tacca, F. Maglia, and G. Spinolo, Spark plasma Sintering and Characterization of Bulk Nanostructured Fully Stabilized Zirconia: Part I. Densification Studies, *J. Mater. Res.*, **19**, 3255-3262 (2004).
[21]U. Anselmi-Tamburini, J.E. Garay, and Z.A. Munir, Fast Low-Temperature Consolidation of Bulk Nanometric Ceramic Materials, *Scripta Mater.*, **54**, 823-828 (2005).
[22]S.R. Casolco, J. Xu, and J.E. Garay, Transparent/Translucent Polycrystalline Nanostructured Yttria Stabilized Zirconia with Varying Colors, *Scripta Mater.*, **58**, 516-519 (2008).
[23]K. Morita, K. Hiraga, B.-N. Kim, and Y. Sakka, High-Strain-Rate Superplasticity in Y_2O_3-Stabilized Tetragonal ZrO_2 Dispersed with 30 vol% $MgAl_2O_4$ Spinel, *J. Am. Ceram. Soc.*, **85**, 1900-1902 (2002).

DENSIFICATION ENHANCEMENT OF ALUMINA BY SANDWICH PROCESS DESIGN

Osayande L. Ighodaro, Okenwa I. Okoli, Ben Wang.

Industrial and Manufacturing Engineering Department, High Performance Materials Institute, FAMU—FSU College of Engineering, 2525 Pottsdamar Street, Tallahassee, Florida 32310

ABSTRACT

Nano-particles densify faster than coarse (or submicron) particles. They also undergo higher volumetric shrinkage than coarse particles during sintering. By fabricating layered structure of coarse and nano particles of alumina powder having an interface of coarse particle/nano-particles, the volumetric shrinkage differential between the distinct layers constitute pressure on the coarse particles while sintering is in progress. The resulting dynamic compressive stress generated promotes mobility of the coarse particles, thus enhancing densification. By grinding off the outer tensile nano-particle layers after densification, the inner coarse particle layer is freed from residual stresses. Comparing the density obtained by this technique with the conventional pressureless sintering, significant enhancement of densification has been achieved, and it is shown that the higher the ratio of the thickness of the fine layer to the thickness of the coarse layer (the thickness ratio), the higher the densification enhancement.

INTRODUCTION

Efforts are being made to improve the processing technologies of ceramic materials in order to improve their mechanical resistance [1 – 3], and densification [4 – 6], especially for those required for structural applications. Improved densification also enhances fracture strength. These informed the use of sintering aids [6 – 10], hot pressing [11, 12] spark plasma sintering (SPS) [13 – 16], and hot isostatic pressing [17] to enhance densification at lower sintering temperatures. Sintering aids are effective for densification enhancement but their applications are not general, as specific aids are effective for specific material(s), and some aids are harmful to other properties of the sintered matrix [7, 10]. Hot pressing and spark plasma sintering processes are costly, involving expensive equipment. The effectiveness of hot pressing and SPS lie in the sustenance of pressure during the sintering process, helping to bring the powder particles together as heat is being applied, thus promoting densification. Nano particles densify faster than coarse (or submicron) particles. They also undergo higher shrinkage than coarse particles during sintering. By fabricating layered structure of coarse and nano particles having an interface of coarse particle /nanoparticles, higher shrinkage rate in the nanoparticle layer(s) constitute pressure on the coarse particles while sintering is in progress. This is similar to hot pressing, though the pressure is not axial in this case. The aim of this work is to employ nanoparticle alumina powder to enhance densification of coarse alumina powder by layered architecture. The nano layer will be subsequently eliminated by polishing to characterize the coarse dense body and evaluate the effectiveness of this technique.

THEORY

Figure 1 shows a schematic of a green compact comprising of coarse ceramic powder sandwiched between fine ceramic powder layers. As sintering commences, the fine layer begins to shrink, thereby exerting stress on the coarse particles. Given that the fine particles

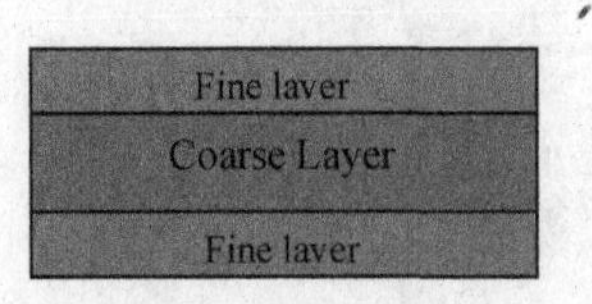

Fig. 1. Schematic of sandwiched coarse powder between nano layers

Fig. 2. Schematic of forces acting on the sandwich structure during sintering

undergo more shrinkage than the coarse particles, then the stress (σ) acting on the coarse particle layer at any time could be represented as $\sigma = E\Delta\varepsilon$, where E is the modulus of the coarse layer and $\Delta\varepsilon$ is the instantaneous difference in shrinkage between the fine particles and coarse particles. Figure 2 shows the forces acting on the structure as sintering commences. The fine particle layers F_A are subjected to tensile stress due to their higher shrinkage while the coarse particle layer (F_B) is subjected to compressive force. These forces are transmitted along the interfaces.

Considering a situation where the different layers are sintered separately, the fine particle layer (A) would shrink to a length L_A, the coarse layer (B) shrinks to a length L_B. But for the sandwich structure (F), a final uniform length L_f, will be attained. This is depicted in Figure 3.

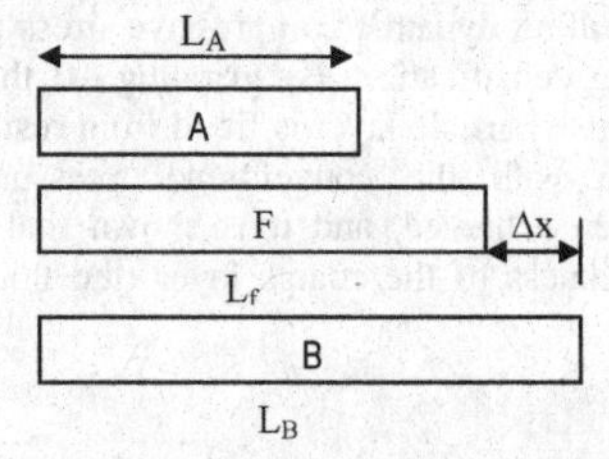

Fig. 3. Schematic of lengths of the different layers

ANALYSIS

The following assumptions are implied:

1. The layers behave linearly elastic
2. The interface is strong and intact throughout

For equilibrium of forces at the interface (from Fig. 2),

$$2F_A = F_B \quad (1)$$

From $F = k\varepsilon$, where F is the active force along the interface, k is the stiffness ($k = EA/L$) and ε is the change in length,

$$2k_A\varepsilon_A = k_B\varepsilon_B \quad (2)$$

$$2E_A A_A(L_f - L_A)/L_A = E_B A_B(L_B - L_f)/L_B \quad (3)$$

Let $E_A = \phi E_B$. Then,

$$\frac{2\phi A_A}{L_A}\left(L_f - L_A\right) = \frac{A_B}{L_B}\left(L_B - L_f\right) \qquad (4)$$

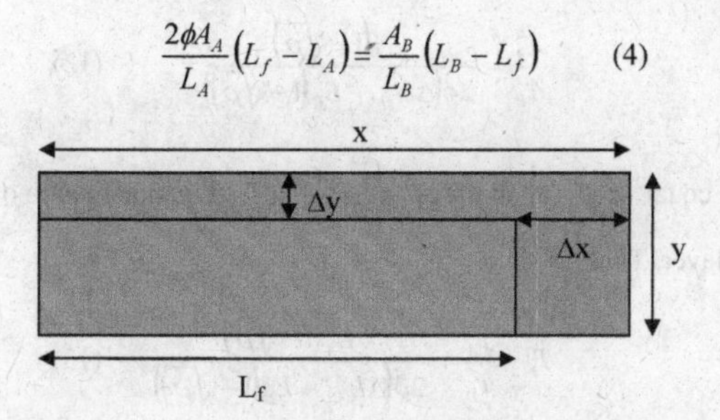

Fig. 4 Schematic of aerial view of plate B with and without sandwich design

Figure 4 shows the aerial view of plate B. Without the fine-particle layers, B is porous, having density ρ%, and dimensions of width y and length x ($x = L_B$). By the sandwich arrangement, the length now shrinks by Δx to the final length (L_f). Simultaneously, the width shrinks by Δy, to bring the piece to full densification. Then equation holds.

$$y\Delta x + x\Delta y - \Delta x\Delta y = xy(1-\rho) \qquad (5)$$

Let $y = nx$, then, $\Delta y = n\Delta x$, from which

$$y\Delta x + nx\Delta x - n(\Delta x)^2 = xy(1-\rho) \qquad (6)$$

$$n(\Delta x)^2 - (y + nx)\Delta x + xy(1-\rho) = 0 \qquad (7)$$

$$n(\Delta x)^2 - 2y\Delta x + xy(1-\rho) = 0 \qquad (8)$$

$$\Delta x = \frac{2y \pm \sqrt{4y^2 - 4nxy(1-\rho)}}{2n} \qquad (9)$$

$$\Delta x = x\left(1-\sqrt{\rho}\right) \qquad (10)$$

Now, $x = L_B$, $\quad L_B - L_f = \Delta x$

$L_f - L_A = L_B - L_A - \Delta x = \Delta L_{AB} - \Delta x$ (See Fig. 3).

Substituting these into (4) gives

$$\frac{2\phi A_A}{L_A}\left(\Delta L_{AB} - L_B\left(1-\sqrt{\rho}\right)\right) = \frac{A_B}{L_B}L_B\left(1-\sqrt{\rho}\right) \qquad (11)$$

$$\frac{A_A}{A_B} = \frac{L_A\left(1-\sqrt{\rho}\right)}{2\phi\left(\Delta L_{AB} - L_B\left(1-\sqrt{\rho}\right)\right)} \qquad (12)$$

Since the layers have equal widths at all times, $\frac{A_A}{A_B} = \frac{t_A}{t_B} = T_r$ = thickness ratio of the surface fine layer to the middle coarse layer. Thus,

$$T_r = \frac{t_A}{t_B} = \frac{L_A\left(1-\sqrt{\rho}\right)}{2\phi\left(\Delta L_{AB} - L_B\left(1-\sqrt{\rho}\right)\right)} \qquad (13)$$

T_r is the thickness ratio of the fine outer layer to coarse inner layer required to fully densify the inner layer. If this ratio is not attained or other parameters such as ΔL_{AB} is not sufficient, ρ is too low (i.e. porosity of plate is too high), or the prevailing temperature is not high enough, then the specimen cannot be densified fully. For enhanced densification up to γ, following similar procedure, (5) becomes:

$$y\Delta x + x\Delta y - \Delta x\Delta y = xy(\gamma - \rho) \qquad (14)$$

$$T_r = \frac{L_A\left(1-\sqrt{1-\gamma+\rho}\right)}{2\phi\left(\Delta L_{AB} - L_B\left(1-\sqrt{1-\gamma+\rho}\right)\right)} \qquad (15)$$

$$\gamma = 1 + \rho - \left[1 - \frac{2\phi T_r \Delta L_{AB}}{L_A + 2\phi T_r L_B}\right]^2 \quad (\rho \le \gamma \le 1) \qquad (16)$$

(16) portrays the maximum densification, γ%, that can be achieved given that ρ% would normally be achieved without this sandwiched design. The parameter ϕ is dynamic; varying from the green state through to the sintered state. In the green state and in the early densification stage its value is greater than unity ($\phi > 1$) since the elastic modulus of the nanoparticles compact is higher than that for the coarse particles compact (also, densification of nanoparticles commence earlier). This is so due to the higher binding strength of the green nanoparticles, making it stiffer. In the course of sintering and as densification progresses, the stiffness of the coarse-layer (E_B) becomes greater than the stiffness of the nano-layer (E_A), as a result of its lower level of porosity. Thus, from some point in the sintering history to the fully sintered state the parameter ϕ will now be less than unity ($\phi < 1$). An average value of ϕ would normally be used for the entire sintering process. To determine ϕ, values of E_A and E_B would be measured during the sintering histories of specimens A and B respectively, under similar conditions of sintering specimens C_i, and corresponding values of ϕ can be determined. The value of ϕ in (16) is then the average of the ϕ values.

EXPERIMENTATION

A key requirement for this process is strong interface between the layers. In order to achieve strong interface between coarse and fine powder a blend of nanopowder and coarse powder was made. 99.97 % pure alumina powder (APS: 150nm, Accumet materials) was blended with 99.98 % pure nanoparticle alumina powder (APS: 10 – 20nm, predominantly γ – phase, 5 – 20% α – phase, Alfa Aesar) in the ratio 1:1. Green compacts were made from this blend by uniaxial compaction at 145 MPa (Specimen B). Next, different masses of this powder were sandwiched between layers of the nanopowder and uniaxially compacted at 145 MPa (specimen C_i, i =1, 2, 3, 4). The thickness of the nanopowder layer was the same for all specimens. The nanopowder was also compacted at the same pressure (specimen A). All compacts were pressureless sintered at 1500 ˚C in air, and allowed to cool in the furnace. Table I shows the thicknesses of the different layers of the specimens. The dimensions of specimen A (L_A) and specimen B (L_B) were measured. The specimens were then polished. During polishing, all regions containing the nanopowder layers (on the surfaces) were completely polished out. The densities were then measured using Archimedes Principles, and microstructures of the surfaces were observed under SEM.

Table I Specimens and their thicknesses

Specimens	Thickness of A t_A (mm)	Thickness of B t_B (mm)	$t_A / t_B = (T_r)$
A	1.25	~	~
B	~	~	~
C_1	1.25	2.0	0.63
C_2	1.25	2.8	0.45
C_3	1.25	3.7	0.34
C_4	1.25	4.1	0.230

RESULTS AND DISCUSSIONS

The materials involved in the experiments are chemically identical, differing only in particle size and porosity. The values of ϕ would change from $\phi > 1$ to $\phi < 1$ from the green state to the sintered state respectively. This brings the average value of ϕ close to unity. More so, since specimen B is composed of the coarse and nano powders in ratio 1;1, the difference between E_A and E_B is further reduced, bringing ϕ closer to unity. Thus, $\phi = 1$ has been adopted as a good approximation in (16) in this work. Measured parameters are: L_A = 42.0mm; L_B, = 48.8mm; density of B = 80.94% (ρ = 0.8094); ΔL_{AB} = 6.80mm. Table 2 shows the parameters for specimens C_i.

Table II Measured densities for different thickness ratios

Specimens	$t_A / t_B = (T_r)$	Density (γ %)
C_1	0.63	96.15
C_2	0.44	91.16
C_3	0.34	89.74
C_4	0.3	85.38
B	0	80.94

Figure 5 is a plot of (13) for given L_A, L_B and ΔL_{AB}. It is seen that the thickness ratio T_r required to fully densify the specimen diminishes exponentially as the density (ρ) (achievable without the sandwich process) increases. This is expected, and this process would not be necessary if the powder can be pressureless sintered to full densification

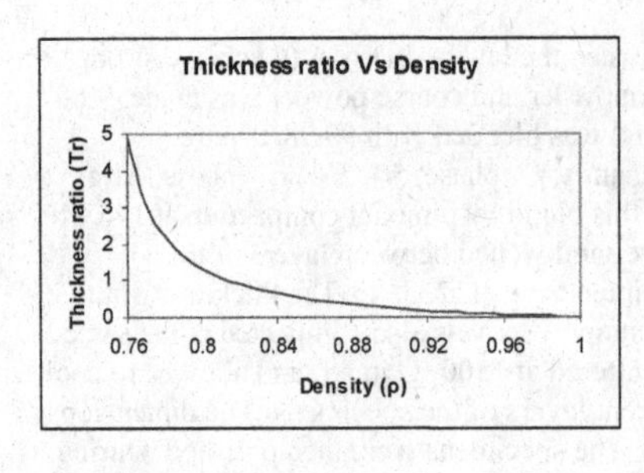

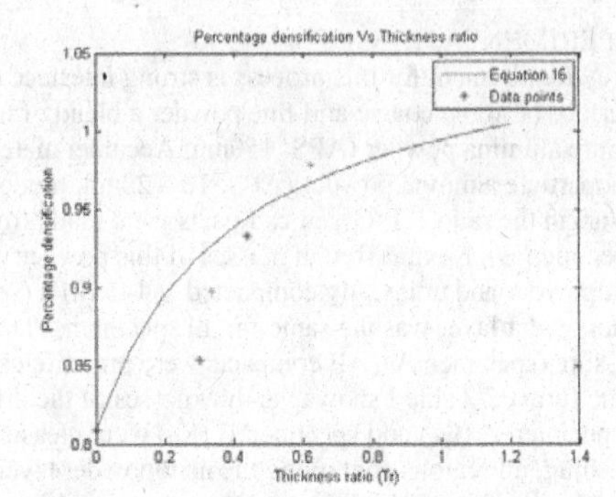

Fig. 5 Plot of thickness ratio required for full densification

Fig.6 Thickness ratio required for a predetermined percentage densification

using the conventional method. Figure 6 shows the thickness ratio required to achieve a predetermined percentage density (γ). (16) and data points are shown. It is seen that densification is enhanced as thickness ratio increases.

Figs. 6a shows the SEM of a sample sintered without using the sandwich method. Figs. 6 (b and c) show the SEM of specimens C_2 and C_1 respectively. Clearly, the micrographs of the sandwiched specimens show less porosity and as such the specimens are more densified. Also, the microstructure of specimen B (Fig. 6a) shows much disunity between the large and nanoparticle contents. This is due to agglomeration of the nanoparticles which form rigid networks with associated porosities and hinders densification [4]. But by the sandwich technique, the dynamic compressive stress generated during sintering breaks the network structure, reduces the agglomeration and increases interparticle contact, thereby enhancing densification. It can be seen from Figs. 6b and 6c that nanoparticles agglomeration is not evident.

Fig. 6a SEM of specimen B

Fig. 6b SEM of specimen C_2

Fig. 6c SEM of specimen C_1

During sintering various mechanisms come to play and temperature and pressure are inevitable requirements. The model presented here is based on the influence of the compacting stress only, during densification. It is intended to show a qualitative effect of harnessing materials having different shrinkages during sintering. Although the data points deviate significantly from the model equation, the trend is maintained; the higher the thickness ratio, the more enhancement in densification. This is expected. The outer layer which generates the compressive stress is also subjected to tensile stress. The thicker this layer, the more the stress it can accommodate before failure and also the higher the pressure it can exert on the inner layer.

SUMMARY

Pressure enhances densification during sintering. During pressureless sintering, dynamic pressure could be generated by the sintering process if materials having different shrinkages are compacted in a sandwich arrangement. The difference in shrinkage would generate some pressure which has the effect of moving the particles closer, thereby increasing interparticle contact and closing up porosities. It has been shown that the higher the thickness ratio, the more the enhancement of densification. Thus this technique is highly potent for pressureless sintering of ceramics at relatively lower compaction pressures and could serve as a substitute for hot pressing.

REFERENCES

1. Y. Fu, Y.W. Gu, H. Du, "SiC Whisker Toughened Al_2O_3-(Ti, W)C Ceramic Matrix Composites," Scripta mater. 44 (2001) 111–116

2. D. Kovar, S. J. Bennison, M. J. Readey, "Crack Stability and Strength Variability in Alumina Ceramics with Rising Toughness – Curve Behavior," Acta mater. 48 (2000) 565 – 578

3. O. L. Ighodaro and O. I. Okoli, "Fracture Toughness Enhancement for Alumina Systems: A Review," Int. J. Appl. Ceram. Technol., 5 [3] 313–323 (2008)

4. B.G. Ravi, Ř. Chaim, A. Gedanken, "Sintering of Bimodal Alumina Powder Mixtures With A Nanocrystalline Component," Nanostructured Materials, Vol. 11, No. 7, Pp. 853–859, 1999

5. A. R. Olszyna, P. Marchlewski, K. J. Kurzydlowski, "Sintering of High-Density, High-Purity Alumina Ceramics," Ceramics International 23 (1991) 323-328

6. J. Wang, S.Y. Lim, S.C. Ng, C.H. Chew, L.M. Gan, "Dramatic effect of a small amount of MgO addition on the sintering of Al,O,-5 ~01% SIC nanocomposite," Materials Letters 33 (1998) 273-277

7. S. Kumar C. Pillai, B. Baron, M. J. Pomeroy, S. Hampshire, "Effect of oxide dopants on densification, microstructure and mechanical properties of alumina-silicon carbide nanocomposite ceramics prepared by pressureless sintering," Journal of the European Ceramic Society 24 (2004) 3317–3326

8. H. Erkalfa, Z. Mlslrh, T. Baykara, "Effect of additives on the densification and microstructural development of low-grade alumina powders," Journal of Materials Processing Technology 62 (1996) 108 115

9. L. Radonjic, V. SrdiC, "Effiect of magnesia on the densification behavior and grain growth of nucleated gel alumina," Materials Chemistry and Physics 47 (1997) 78-84

10. Y. Haitao, G. Ling, Y. Runzhang, Y. Guotao, H. Peiyun, "Effect of MgO/CeO2 on pressureless sintering of silicon nitride," Materials Chemistry and Physics 69 (2001) 281–283

11. H. Tomaszewski, M. Boniecki, H. Weglarz, "Toughness-curve behaviour of alumina-SiC and ZTA-SiC composites," Journal of the European Ceramic Society 20 (2000) 1215 – 1224

12. M. Aldridge and J. A. Yeomans, "The Thermal Shock Behavior of Ductile Particle Toughened Alumina Composites," Journal of the European Ceramic Society 19 (1998) 1769 – 1775

13. Gao, H. Z. Wang, J. S. Hong,a H. Miyamoto, K. Miyamoto, Y. Nishikawa and S. D. D. L. Torre, "Mechanical Properties and Microstructure of Nano-SiC/Al_2O_3 Composites Densified by Spark Plasma Sintering," Journal of the European Ceramic Society 19 (1999) 609 – 613

14. G. Bernard-Granger, C. Guizard, "Spark plasma sintering of a commercially available granulated zirconia powder: I. Sintering path and hypotheses about the mechanism(s) controlling densification," Acta Materialia 55 (2007) 3493–3504

15. B. Kim,K. Hiraga, K. Morita and H. Yoshida, "Spark plasma sintering of transparent alumina," Scripta Materialia 57 (2007) 607–610

16. X. Yao, Z. Huang, L. Chena, D. Jianga, S. Tana, D. Michel, G. Wang, L. Mazerolles, J. Pastol, "Alumina–nickel composites densified by spark plasma sintering," Materials Letters 59 (2005) 2314 – 2318

17. J. Echeberria, J. Tarazona, J. Y. He, T. Butler, F. Castro, "Sinter-HIP of a-alumina powders with sub-micron grain sizes Journal of the European Ceramic Society 22 (2002) 1801–1809

PROCESSING FACTORS INVOLVED IN SINTERING β-Si_3N_4-BASED CERAMICS IN AN AIR ATMOSPHERE FURNACE

Kevin P. Plucknett
Materials Engineering Program, Department of Process Engineering and Applied Science, Dalhousie University, 1360 Barrington Street, Halifax, Nova Scotia, B3J 1Z1, CANADA

ABSTRACT

Conventionally, silicon nitride (Si_3N_4) based ceramics are sintered in a controlled N_2 atmosphere furnace in order to prevent their decomposition. However, it has recently been demonstrated that both dense and porous β-Si_3N_4 ceramics can be successfully sintered in a conventional air atmosphere furnace (i.e. with molybdenum disilicide ($MoSi_2$) heating elements), provided that a localized N_2 atmosphere is retained around the samples during sintering. This can be achieved by packing them in a pure Si_3N_4 powder, held within an alumina (Al_2O_3) crucible, and then heating to the sintering temperature at high enough rates to promote oxidation of the surface of the packing powder. Oxide scale formation then seals an N_2-rich environment around the sample, as residual oxygen is consumed by partial Si_3N_4 oxidation; this process can result in a small, net sample weight gain, which results in the formation of a limited amount of silicon oxynitride (Si_2N_2O) in the material during sintering. For the case of combined yttria (Y_2O_3) and Al_2O_3 sintering additions, densities in excess of 98 % of theoretical can be achieved at the highest sintering temperatures (e.g. 1700 to 1750°C). For materials prepared with single Y_2O_3 additions, sintered densities approaching 90 % of theoretical were achieved, which is significantly higher than similar samples sintered in N_2. The use of high heating rates helps to suppress the oxidation weight gain for both dense and porous Si_3N_4. This results in lower sintered densities for the single additive samples, and the elimination of Si_2N_2O from the sintered product. The influence of heating rate, crucible packing arrangement, sintering additive composition and packing powder type are discussed.

INTRODUCTION

The development of new methods to reduce the processing costs of advanced ceramics are highly desirable, such that materials may be implemented into new application areas were their performance characteristics are favorable but their production costs are currently prohibitive. One such material is silicon nitride (Si_3N_4), which possesses an extremely useful combination of physical, chemical and thermal properties.[1] In particular, ceramics based on the higher temperature, β-Si_3N_4 form can exhibit an excellent combination of strength and toughness, while being of low mass (density typically between 3.2 and 3.4 g/cm^3), and possessing high corrosion and oxidation resistance, together with excellent wear resistance. For example, combined flexure strength and toughness values of 1100 MPa and 11 MPa.m$^{1/2}$ have been reported for these *in-situ* reinforced materials.[2]

In the present work, a simple method has been developed to potentially lower the processing costs of Si_3N_4-based ceramics, through sintering with conventional air atmosphere furnaces rather than expensive controlled atmosphere (i.e. N_2) systems. The approach that has been taken involves enclosing samples in a powder bed of pure α-Si_3N_4, which is itself held in an Al_2O_3 crucible.[3,4] During the initial heating ramp to the sintering temperature, the outer surface of the α-Si_3N_4 powder rapidly oxidizes, resulting in the samples being contained in a nominally sealed environment. Sintering then occurs in a predominantly N_2-rich atmosphere, helping to avoid excessive oxidation of the samples, even at temperatures as high as 1750°C.[3,4] This approach allows the preparation of multi-additive Si_3N_4 ceramics to densities in excess of 98 % of theoretical, while retaining a comparable microstructure to similar compositions processed in N_2. The processing influences of a variety of experimental variables, such as furnace heating rate, crucible arrangement and α-Si_3N_4 powder type are presented, with reference to both high density and porous β-Si_3N_4 ceramics.

EXPERIMENTAL PROCEDURES

Two α-Si_3N_4 powders have been used in the present work, namely Starck LC12-SX and Ube SN E-10, which are both sub-micron and predominantly α-Si_3N_4. In order to prepare porous β-Si_3N_4 ceramics, a 5 wt. % Y_2O_3 sintering addition was employed, which is designed to retard densification while still promoting the α- to β-Si_3N_4 transformation. For the preparation of dense samples, a combination of Y_2O_3 and Al_2O_3 additions were used, with a composition of either 12 wt. % Y_2O_3/3 wt. % Al_2O_3 or 8 wt. % Y_2O_3/2 wt. % Al_2O_3; these compositions are subsequently referred to as 12/3 and 8/2 respectively. Samples were prepared by ball milling the appropriate powders in propan-2-ol (isopropyl alcohol) for 24 hours, using 10 mm diameter zirconia media. After milling the powders were dried and then crushed using a pestle and mortar, or sieved through a -75 μm mesh sieve. Pellets were pressed uniaxially at ~30 MPa, and then isostatically at either 170 or 200 MPa. For air sintering, the pellets were sited in a powder bed of the identical α-Si_3N_4 powder as used in the samples themselves. Three different packing arrangements were used, as shown in Figure 1.

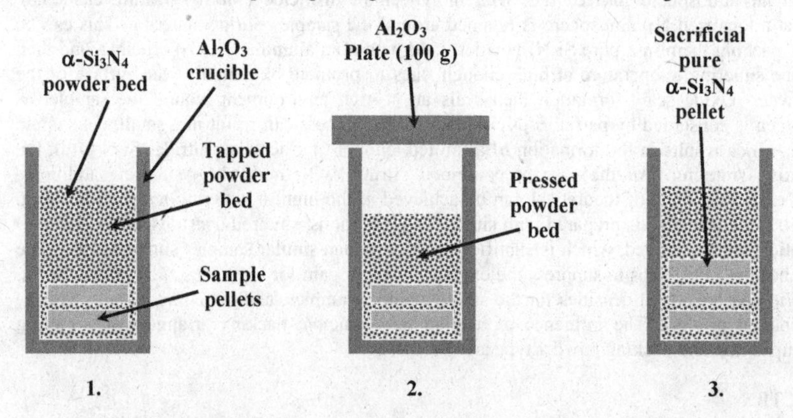

Figure 1. Schematic representations of the three different sample/crucible arrangements (subsequently referred to as CA1-3) used in the preparation of porous and dense Si_3N_4 ceramics in air.

Sintering was performed using conventional $MoSi_2$ element furnaces, and the nominally maximum achievable heating rate was employed for each furnace to minimize potential sample oxidation. As a consequence, two different sample heating rate cycles were assessed, which are shown schematically in Figure 2 (subsequently referred to as HR1 and HR2). After sintering, the densities of multi-additive samples were measured by immersion in water, while those of the porous single additive samples were measured geometrically. Scanning electron microscopy (SEM; Model S-4700, Hitachi Industries, Tokyo, Japan) was performed on carbon-coated surfaces, for both polished samples and fracture surfaces. Crystalline phase identification was conducted using X-ray diffraction (XRD; Model D-8 Advance, Bruker AXS, Inc., Madison, WI), with the α- to β-Si_3N_4 ratio calculated using the relationship of Yeheskel and Gefen.[5]

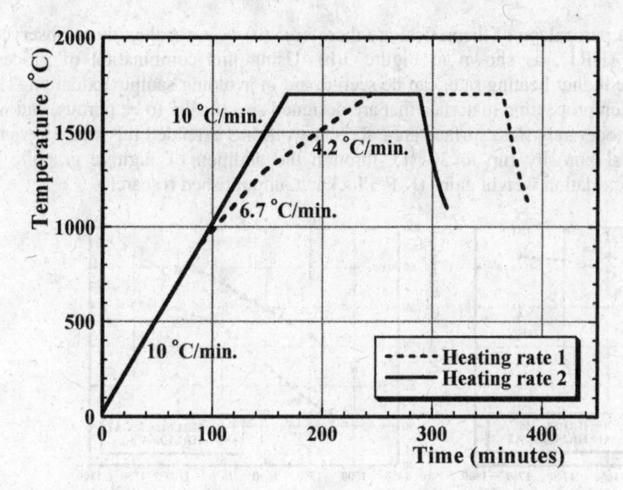

Figure 2. The two sintering cycles used for processing Si_3N_4 ceramics in air (subsequently referred to as HR1 and 2).

RESULTS AND DISCUSSION

Porous β-Si_3N_4 ceramics

After sintering, samples prepared in air are essentially white in appearance, and show no sign of surface oxidation, possessing a highly smooth surface and good dimensional stability. The influence of sintering temperature on the density of porous Si_3N_4 ceramics prepared with single Y_2O_3 sintering additions is shown in Figure 3(a). Using the processing combination of HR1/Starck α-Si_3N_4/CR1, it can be seen that a maximum density of ~2.8 g/cm^3 can be achieved when sintering at 1700℃. In this instance sintering is accompanied by significant weight gains, up to ~2.5 wt. % for the 5 wt. % Y_2O_3 additive content material. This weight gain is a result of sample oxidation, such that SiO_2 is formed following:

$$Si_3N_{4(s)} + 3O_{2(g)} \rightarrow 3SiO_{2(s,l)} + 2N_{2(g)} \quad (1)$$

Ultimately, during sintering above the liquid formation temperature, which is ~1550°C for the Si-Y-O-N system,[6] the presence of excess SiO_2 would be expected to result in the formation of Si_2N_2O in the final product, following:

$$Si_3N_{4(s)} + SiO_{2(s,l)} \rightarrow 2Si_2N_2O_{(s)} \quad (2)$$

As a consequence, the theoretical density of the sintered samples therefore needs to be modified, depending on the measured weight gain. In this instance, a theoretical density of ~2.85 g/cm^3 is used for Si_2N_2O,[7] compared with 3.19 g/cm^3 for β-Si_3N_4.[8] The percentage of theoretical density achieved is shown in Figure 3(b).

In contrast, using a combination of HR2/Ube α-Si_3N_4/CA3, the sintered densities are slightly lower (Figure 3(a)). In this instance, the higher heating rate appears sufficient to essentially prevent any oxidation of the samples, such that they show slight weight losses, rather than weight gains, with the highest weight loss of up to 2 wt. % noted for samples sintered at 1750℃ (it is notable that similar compositions sintered in a N_2 atmosphere, in an α-Si_3N_4/boron nitride powder bed within a graphite crucible, showed weight losses of up to ~6 wt. % at the same sintering temperature). As a consequence, the sintered densities as a percentage of theoretical are therefore ~10 % lower than those observed for the slower heating rate (HR1), as shown in Figure 3(b). Using this combination of processing parameters, especially the higher heating rate, can be seen to aid in avoiding sample oxidation. This is particularly important when processing materials that are designed specifically to be porous, and which consequently possess a moderately high surface area. Recent work has extended this study to samples with even higher residual porosity (up to 30 %), through the addition of fugitive graphite filler particles, again avoiding oxidation weight gains (K.P. Plucknett, unpublished research).

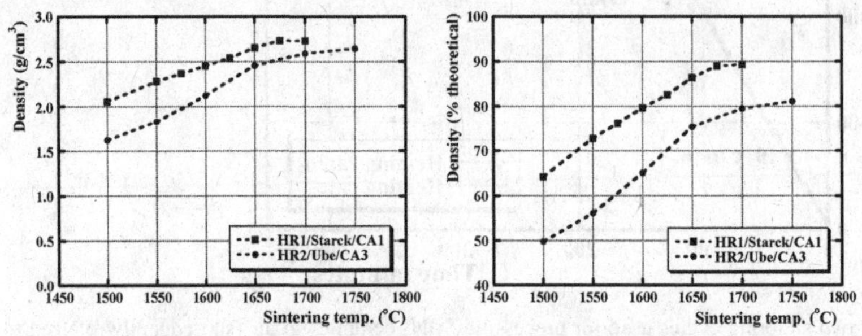

Figure 3. (a) The measured density of 5 wt. % Y_2O_3 samples prepared in air using two different processing cycles. (b) Data from (a) re-plotted as the percentage of theoretical density as a function of sintering temperature; the theoretical densities for samples prepared using HR1/Starck α-Si_3N_4/CA1 are adjusted to account for Si_2N_2O formation, due to a pronounced oxidation weight gain. Adapted in part from ref. [4].

It is notable that the sintered densities achieved are somewhat higher than comparable materials sintered in a pure N_2 atmosphere where, typically, sintered densities in the range of 60 % of theoretical are achieved.[9,10] For the case of HR1/Starck α-Si_3N_4/CA1, this is in part due to the presence of a thin, viscous SiO_2 surface layer formed through oxidation, which contributes a viscous-flow component to sintering.[4] For HR2/Ube α-Si_3N_4/CA3, it is likely that higher densities result as weight loss reactions, typical in a N_2 atmosphere, are suppressed.

As noted above, the use of a lower heating rate sintering cycle (HR1) can be anticipated to result in the formation of Si_2N_2O, following reactions (1) and (2). This is confirmed for both the surface and bulk regions of these samples, as shown in Figure 4. In particular, it is clear that the surface of these samples is predominantly Si_2N_2O, and that a gradient in Si_2N_2O concentration exists into the bulk region of the material.[4] In contrast, the surface of the samples sintered at the higher heating rate show no evidence of Si_2N_2O formation (Figure 5 (a)), confirming the benefits of suppressing the weight gain phenomenon (Equation (1)).

In this instance, the surface α- to β-Si_3N_4 transformation behavior is broadly similar to that for the materials processed at the lower overall heating rate (Figure 5(b)). However, it is clear that the extent of α- to β-Si_3N_4 transformation at each specific sintering temperature is retarded when using the higher heating rate (HR2). This is understandable, based on Figure 2, in that the time to reach the actual sintering hold can be extended by ~75 minutes when heating to 1700°C for the lower heating rate (HR1). As a consequence, there is less time available for sintering to occur using HR2 (i.e. the time that the samples are exposed to temperatures in excess of 1550°C, where a second phase Si-Y-O-N liquid can be anticipated to form[6]). Critically, the extended heating cycle (HR1) exposes the samples to a prolonged period between initial surface oxidation of the protective powder bed, estimated to be

~1000ºC, and the final sintering hold temperature (e.g. 1700ºC), which is believed to be the reason for the significantly higher extent of oxidation in the samples heated at the lower rate.

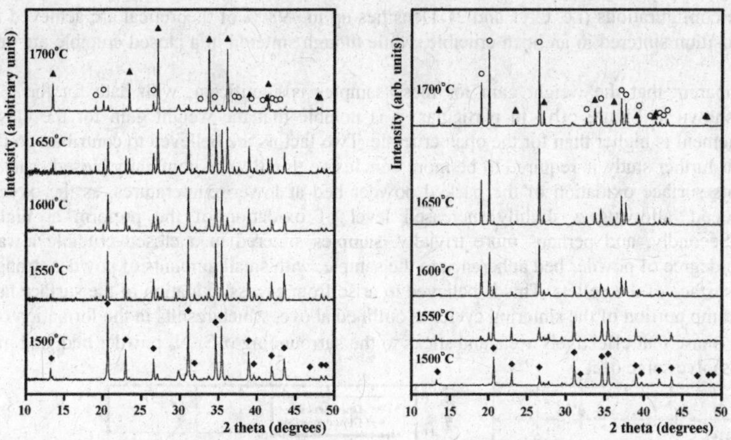

Figure 4. The effects of sintering temperature on (a) bulk and (b) surface phase evolution for 5 wt. % Y_2O_3 samples prepared in air using HR1/Starck α-Si_3N_4/CA1, determined using XRD (◆:α-Si_3N_4; ▲:β-Si_3N_4; O:Si_2N_2O). Adapted from ref. [4].

The microstructure of the porous samples sintered at 1700 and 1750ºC are similar to materials processed in N_2, with a network of anisotropic β-Si_3N_4 grains combined with a continuous, interconnected network of porosity. At the highest temperatures, grain sizes were typically of the order of 3-5 μm in length, with aspect ratios of ~5:1.[4] The formation of a Si_2N_2O-rich surface zone, when using HR1, resulted in some surface coarsening of the grain structure to a depth of ~50 μm, and the formation of a thin, liquid-like (i.e. SiO_2) residue on the outer surface.[4]

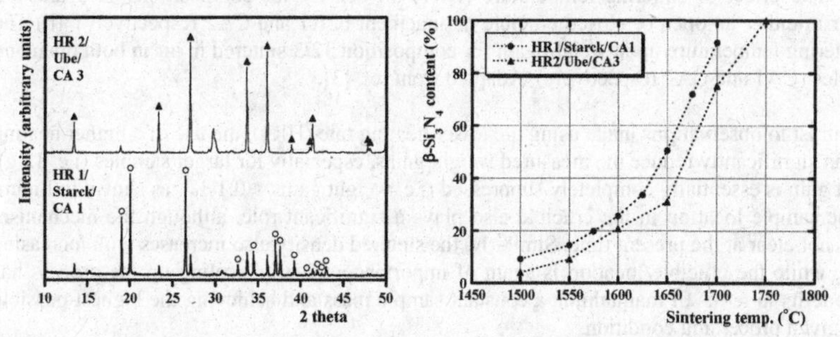

Figure 5. (a) Comparison of the surface phase composition for samples sintered at 1700 and 1750ºC, for HR1 and HR2 respectively, demonstrating the absence of Si_2N_2O formation when using the higher heating rate. (▲:β-Si_3N_4; O:Si_2N_2O). (b) The volume fraction of β-Si_3N_4, relative to α-Si_3N_4, formed at the sample surface (ignoring Si_2N_2O content), as a function of sintering temperature for the two primary processing approaches. Adapted in part from ref. [4].

High-density β-Si_3N_4 ceramics

The effects of sintering temperature on density for both the 12/3 and 8/2 compositions, prepared with the Starck α-Si_3N_4 powder and a slow heating rate (HR1), are shown in Figure 6(a) for both open and closed crucible configurations (i.e. CA1 and 2). Densities up to ~98 % of theoretical are achieved for the 12/3 composition sintered in an open crucible, while though sintered in a closed crucible are very slightly lower.

It is apparent that the weight gain for these samples is significant, with data for the 12/3 compositions shown in Figure 6(b). In particular, it is notable that the weight gain for the closed crucible arrangement is higher than for the open crucible. Two factors are believed to contribute to this effect, although further study is required to be more conclusive. Firstly, the nominally closed crucible partially inhibits surface oxidation of the packed powder bed at lower temperatures, as the oxygen supply is reduced, allowing a slightly increased level of oxidation of the preform at higher temperatures. Secondly, and perhaps more trivially, samples sintered in a closed crucible always exhibited some degree of powder bed adherence to the sample, with small amounts of powder strongly bonded to the surface of the pellets. This is believed to arise from excess oxidation of the surface later in the heating ramp portion of the sintering cycle, as outlined above, which results in the formation of a residual glassy phase that effectively wets and sticks to the surrounding α-Si_3N_4 powder bed, and may even start to dissolve some of it.

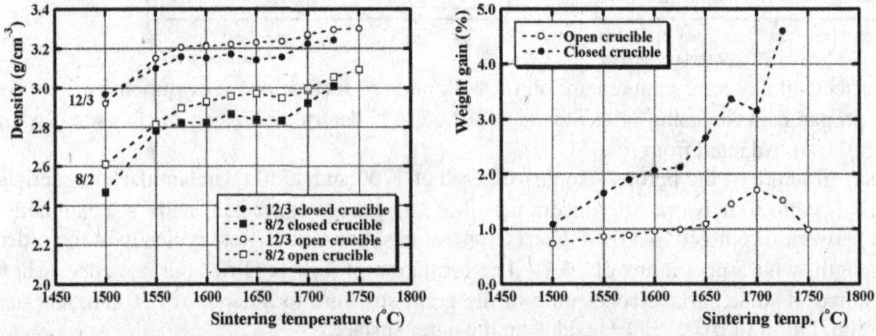

Figure 6. (a) The effect of sintering temperature (HR1) on density for compositions 12/3 and 8/2 sintered in air in either an open or closed crucible arrangement (CA1 and CA2 respectively). (b) The effects of sintering temperature upon weight gain for composition 12/3 sintered in air in both open and closed crucibles (CA1 and CA2 respectively). Adapted from ref. [3].

In contrast to observations made using the lower heating rate (HR1), the use of a higher heating rate (HR2) can significantly reduce the measured weight gains, especially for larger samples (i.e. 12 g), where weight gain is essentially completely suppressed (i.e. weight gains < 0.1 %), as shown in Figure 7(a). Here the sample location in the crucible also plays a significant role, although the mechanism behind this is not clear at the present time. Similarly, the sintered density also increases with increasing sample mass, while the crucible location is again of importance. Clearly, scaling up the process has significant benefits in terms of maintaining a constant sample mass and achieving the highest possible density for a given processing condition.

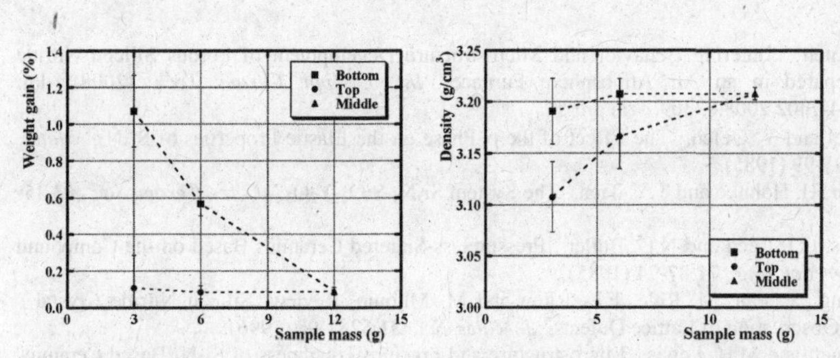

Figure 7. (a) The effects of sample mass and crucible location on weight gain for composition 12/3, prepared with Ube SN E-10 α-Si_3N_4, and using HR2 and CA3. (b) The effects of sample mass and crucible location on sintered density for composition 12/3, prepared with Ube SN E-10 α-Si_3N_4, and using HR2 and CA3. In both cases, "Top" and "Bottom" relate to two samples sited below a sacrificial, pure α-Si_3N_4 pellet, while "Middle" relates to a single sample below a sacrificial pellet.

CONCLUSIONS

Dense and porous β-Si_3N_4 ceramics have been prepared by sintering in an air atmosphere furnace. Sintering in air is generally similar to processing in a N_2 atmosphere, resulting in comparable microstructures and α- to β-Si_3N_4 transformation behavior. Lower heating rates result in sample mass increase, due to oxidation, and an increase in sintered density. However, excess oxidation results in the formation of Si_2N_2O, through the reaction of Si_3N_4 and SiO_2, with higher Si_2N_2O concentrations noted at the sample surface. Higher heating rates significantly reduce sample oxidation, such that Si_2N_2O formation can be totally eliminated. This results in the retention of higher levels of porosity in the samples that are specifically designed to be porous. Increasing the sample mass also decreases the measured weight gain, and results in slightly higher sintered densities. It can therefore be generally recommended that sintering Si_3N_4 ceramics in air should be performed using a high heating rate (i.e. a minimum of 10°C/min) to avoid unwanted reactions occurring throughout the samples that are being sintered. Current work is targeted at thermal and mechanical property evaluation for porous materials prepared with a variety of porosity content (e.g. using fugitive fillers).

ACKNOWLEDGEMENTS

The author would like to thank NSERC for provision of funds through the Discovery Grants program. The support of the Canada Foundation for Innovation, the Atlantic Innovation Fund, and other partners who helped fund the Facilities for Materials Characterisation, managed by the Dalhousie University Institute for Materials Research, are also gratefully acknowledged for access to the FE-SEM. Mr. Mervin Quinlan is also thanked for the provision of data relating to sample mass effects.

REFERENCES

1. G. Petzow and M. Herrmann, 'Silicon Nitride Ceramics,' *Struct. Bond.*, **102** 47-167 (2002).
2. P.F. Becher, E.Y. Sun, K.P. Plucknett, K.B. Alexander, C.-H. Hsueh, H.-T. Lin, E.-S. Kang, K. Hirao and M.E. Brito, 'Microstructural Design of Silicon Nitride with Improved Fracture Toughness, Part I: Effects of Grain Shape and Size,' *J. Am. Ceram. Soc.*, **81** 2821-30 (1998).
3. K.P. Plucknett and H.-T. Lin, 'Sintering Silicon Nitride Ceramics in Air,' *J. Am. Ceram. Soc.*, **88** 3538-41 (2005).

4. K.P. Plucknett, 'Sintering Behavior and Microstructure Development of Porous Silicon Nitride Ceramics Prepared in an Air Atmosphere Furnace,' *Int. J. Appl. Ceram. Tech.* (2008); doi: 10.1111/j.1744-7402.2008.02309.x (in press).
5. O. Yeheskel and Y. Gefen, 'The Effect of the α-Phase on the Elastic Properties of Si_3N_4,' *Mater. Sci. Eng.*, **71** 95-99 (1985).
6. L.J. Gaukler, H. Hohnke and T.Y. Tien, 'The System Si_3N_4-SiO_2-Y_2O_3,' *J. Am. Ceram. Soc.*, **63** 35-37 (1980).
7. M.H. Lewis, C.J. Reed and N.D. Butler, 'Pressureless-Sintered Ceramics Based on the Compound Si_2N_2O,' *Mater. Sci. Eng.*, **71** 87-94 (1985).
8. C.-M. Wang, X. Pan, M. Rühle, F.L. Riley and M. Mitomo, Review: Silicon Nitride Crystal Structure and Observation of Lattice Defects,' *J. Mater. Sci.*, **31** 5281-98 (1996).
9. K.P. Plucknett and M.H. Lewis, 'Microstructure and Fracture Toughness of Si_3N_4 Based Ceramics Subjected to Pre-Sinter Heat Treatments,' *J. Mater. Sci. Lett.*, **17** 1987-90 (1998).
10. K.P. Plucknett, M. Quinlan, L.B. Garrido and L.A. Genova, 'Microstructural Development in Porous β-Si_3N_4 Ceramics Prepared with Low Volume RE_2O_3-MgO-(CaO) Additions (RE = La, Nd, Y, Yb),' *Mater. Sci. Eng. A*, **489** 337-50 (2008).

PRESSURELESS SINTERING OF BORON CARBIDE IN AN Ar ATMOSPHERE CONTAINING GASEOUS METAL SPECIES

Hiroyuki Miyazaki, You Zhou, Hideki Hyuga and Yu-ichi Yoshizawa
National Institute of Advanced Industrial Science and Technology (AIST)
Anagahora 2266-98, Shimo-shidami, Moriyama-ku, Nagoya 463-8560, Japan

Takeshi Kumazawa
Mino Ceramic CO., LTD
1-46, Kamezaki Kitaura-cho, Handa-Shi 475-0027, Japan

ABSTRACT

Boron carbide (B_4C) was densified up to 97.4 % of theoretical density at 2226°C by employing an Ar atmosphere containing gaseous Al and Si species at ambient pressure. X-ray diffraction analysis of the sintered B_4C revealed that both Al and Si elements infiltrated into the green compacts and reacted with B_4C to form SiC and Al_4SiC_4 (m.p. 2037°C) during the sintering. Pockets of the secondary phase observed in the polished and etched surface of the densified sample possessed irregular shapes, indicating the formation of liquid phase during heating.

INTRODUCTION

Boron carbide (B_4C) ceramics is one of the attractive materials for structural components especially where high specific modulus and/or hardness are required since it possesses very high hardness, high Young's modulus and low theoretical densities.[1] However, the strong covalent bonding of B-C hinders pressureless sintering. Densification of B_4C ceramics using hot-pressing limits their product shapes to simple ones or increases the machining cost. Thus, B_4C ceramics has been only used for the special applications such as sand blast nozzles, etc.

Many attempts have been conducted to sinter B_4C at ambient pressure by adding sintering aids such as C,[2-4] Al_2O_3[5,6], TiC[7] and TiB_2.[8] However, a lot of sintering additives are required for full densification, which deteriorates the ideal characteristics of the B_4C. The alternatives to the sintering additives are those techniques which employed a He-H_2 atmosphere[9] or rapid heating rate,[10] both of which are difficult to apply for industrial production. In our previous study, green compacts of B_4C were heated in an Ar atmosphere containing gaseous metal species to densify them without external pressure.[11] It was found that pressureless sintering of B_4C up to 97.4% of theoretical density was achievable at 2226°C with the sintering-aid gases. In this study, crystalline phases in the densified samples were identified using X-ray diffraction (XRD) analysis and their microstructures were observed with optical microscopy and scanning electron microscopy (SEM). The mechanism of pressureless sintering was discussed.

EXPERIMENTAL PROCEDURE

Commercially available B_4C powder (medium particle size: 0.8 μm) was used as a starting powder. The characteristics of the powder such as specific surface area, particle size distribution and impurities were described in our previous paper.[11] XRD analysis of the raw powder in the previous study has shown the existence of minor contents of graphite and boron oxide. The raw powder was uniaxially pressed at 9.8 MPa into pellets of ϕ25 x 5 mm, followed by cold isostatic pressing at 490 MPa. The green compacts and the powder mixture of SiC and aluminum were set in a graphite crucible (Figure 1). The weight ratio of SiC to Al powders was 30 : 1. The compacts were placed on a carbon block so that they did not touch the powder mixture. The crucible was heated using a graphite resistance furnace up to 2104 -2226°C at a heating rate of 10°C/min. The furnace was evacuated by the diffusion pump from R.T. to 1900 °C, followed by Ar-gas inlet to ambient pressure. The soaking time for each target temperature (2104 - 2226°C) was 4 h.

Bulk densities of the samples were measured using the Archimedes method with distilled water. The relative density was obtained by dividing the bulk density by the theoretical density of B_4C of 2.52 g/cm^3. The sintered samples were ground with a diamond wheel followed by polishing with 3μm and 0.5μm diamond slurry to obtain mirror-finish surfaces. The phase analysis was carried out by X-ray diffraction using Cu Kα radiation. The samples were etched in KIO_4-saturated phosphoric acid. The microstructures of the samples were observed with a metallurgical microscope. The surfaces of samples were coated with a 30 nm thick layer of gold using a sputter coater before microstructural observation by scanning electron microscopy.

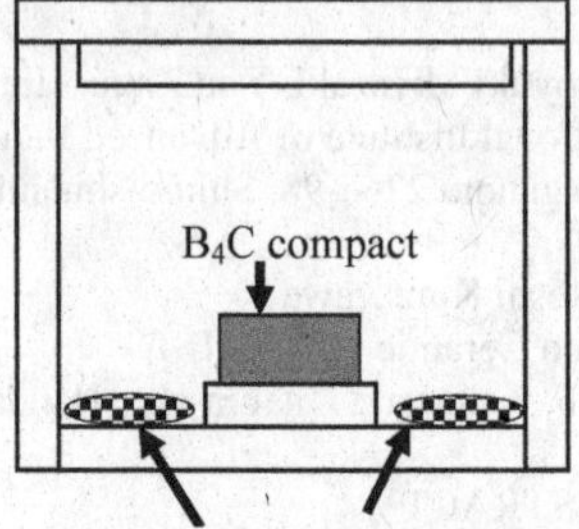

Figure 1. Schematic illustration of the arrangement of the B_4C compact and powder mixture of Al and SiC in a graphite crucible.

RESULTS

Figure 2 shows the relative densities of the B_4C compacts versus temperature. The density of the B_4C compacts in the Ar atmosphere containing gaseous metal species was improved especially at the relatively lower temperature ~2100°C and the maximum relative density of 97.4% was obtained at 2226°C. By contrast, the densification without the sintering-aid gas was difficult unless the compact was heated up to 2226°C.

XRD charts of both samples sintered at 2187°C with and without metallic gas species are presented in figure 3. Only a small amount of graphite was detected in the sample heated without sintering-aid gas, which is derived from the impurities in the raw powder. By contrast, both Al_4SiC_4 and SiC phases were identified besides the graphite phase in the sample sintered with the metallic gas species. It is obvious that Al_4SiC_4 and SiC phases in the latter sample were formed by the reaction of B_4C with both elements of Al and Si which diffused into the green compacts through vapor-phase transport. Boron oxide phase which existed in the raw powder disappeared in the two samples, indicating that B_2O_3 was reduced in the carbon-containing atmosphere.

Figure 4 (a) shows an optical micrograph of the sample heated at 2104°C in the pure Ar gas. Significant neck growth was observed although density of the sample was 84.6 % of the theoretical density. Several small pores with size less than ~ 2 μm were connected to each other to form irregular shapes and were distributed homogeneously in the sample. Over etched sample also revealed that grain coarsening was associated with the small shrinkage (figure 4(b)). When the sample was heated up to 2187°C, the pores exhibited round shape instead of the irregular one (figure 5). These pores were frequently observed in the sample, which was consistent with its low relative density of 94.9%. Significant grain coarsening such as exaggerated grain growth was not observed, which may be attributed to the pores distributed homogeneously in both samples.

By contrast, many isolated small pores in the size of less than ~2 μm were found in the sample sintered at 2104 °C in the Ar atmosphere containing both Al and Si gaseous elements (figure 6). It should be noted that the grain size appeared to be not different considerably from that in the sample heated at the same temperature without the sintering-aid gas (figure 4 (b)) although the shrinkage was enhanced greatly as compared with that of the sample heated in the pure Ar atmosphere.

Figure 7 shows the microstructure of the sample heated at 2187°C in the Ar atmosphere with the sintering-aid gas. Most of the small pores at the triple grain junctions in figure 6 disappeared and only a few of them grew up to ~5 μm after heating at 2187°C. It is obvious that the annihilation of these small pores through grain-boundary diffusion resulted in further densification to 97.1% of theoretical density. Abnormal grain growth took place concurrently and a few pores were trapped in it, which might be obstacles to the full densification of the B_4C even after heating at 2226°C. Figure 8 (a) shows the SEM micrograph of the sample. Small bright secondary phases with irregular shape were often observed at the triple junctions, clearly indicating the presence of liquid phase during the sintering at high temperature. The secondary phases became much brighter when they were observed with a back-scattered electron image (figure 8(b)), which suggested that the constituent elements of the secondary phase were heavier that both B and C. Based on the results of the XRD analysis mentioned above, it was reasonably estimated that the candidate compound for this phase was SiC or Al_4SiC_4. The melting point of Al_4SiC_4 is 2037°C, whereas SiC never melts but sublimates. Consequently, the bright and irregular-shaped phase is likely to be Al_4SiC_4.

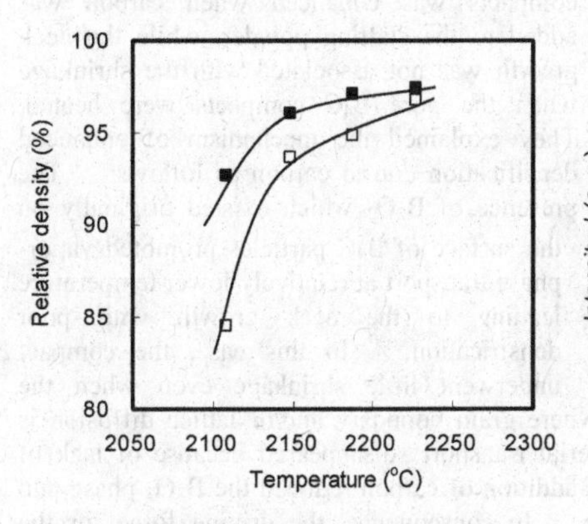

Figure 2. Relative density versus sintering temperature for both B_4C compacts heated in Ar atmospheres with gaseous metal species (solid square) and without gaseous metal species (open square).

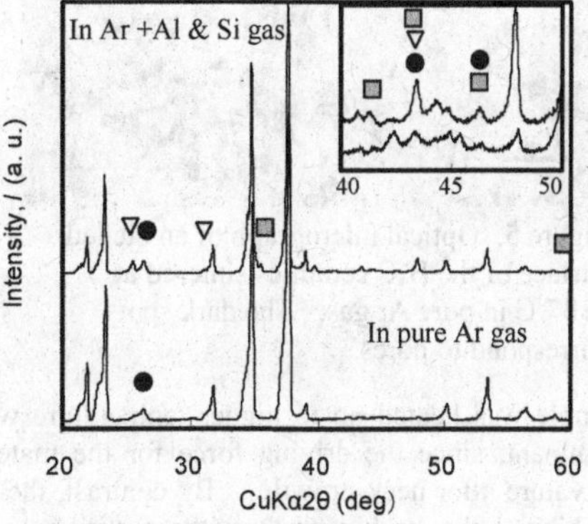

Figure 3. XRD charts of both B_4C samples sintered at 2187°C with and without the sintering-aid gas. ■: SiC, ∇: Al_4SiC_4, ●: graphite.

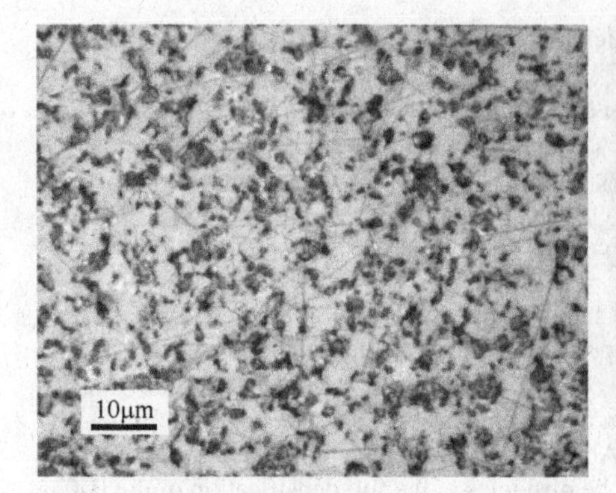

Figure 4 (a). Optical micrograph of a weakly etched surface of the B_4C ceramics sintered at 2104°C in pure Ar gas. The dark spots correspond to pores.

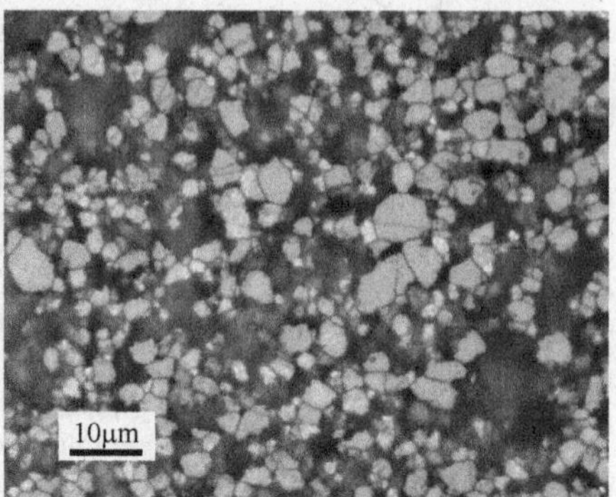

Figure 4 (b). Optical micrograph of a severely etched surface of the B_4C ceramics sintered at 2104°C in pure Ar gas.

Figure 5. Optical micrograph of an etched surface of the B_4C ceramics sintered at 2187°C in pure Ar gas. The dark spots correspond to pores.

DISCUSSION

The mechanism of grain coarsening with negligible shrinkage for B_4C has been well documented by Schwetz and Greliner[3] and Dole et al.[4] They found that the shrinkage of B_4C compacts was enhanced when carbon was added to the starting powder, while the neck growth was not associated with the shrinkage when the pure B_4C compacts were heated. They explained the mechanism of enhanced densification due to carbon as follows: The presence of B_2O_3 which existed originally on the surface of B_4C particles promoted vapor-phase transport at relatively lower temperature, leading to the neck growth with poor densification. In this case, the compact underwent little shrinkage even when the sample was heated up to higher temperature where grain boundary and/or lattice diffusion is dominant, since the driving force for the material transport disappeared because of lack of curvature after neck growth. By contrast, the addition of carbon reduced the B_2O_3 phase and prohibited the neck growth without shrinkage. In consequence, the driving force for the material transfer survived until the sintering temperature reached the region where grain boundary and/or lattice diffusion is predominant. Thus, the sample experienced enough densification at relatively higher temperatures. The same explanation was employed for the improved densification for the samples sintered under the hydrogen atmosphere or with rapid heating rate.[9,10]

It is obvious that the vapor-phase transport was also operating during the sintering of our samples in the pure Ar gas since the significant neck growth was observed irrespective of the little shrinkage. The same analogue to the shrinkage due to the carbon addition seems to be applicable to our case where the densification was improved under the presence of Al and Si gaseous species since Al gas and/or Al compounds gases has a reducing ability at high temperatures just like hydrogen. That is, gaseous Al and Si elements infiltrated into the green body and removed the thin B_2O_3 films on the B_4C particles at the initial stage of the heating, which preserved the driving force for densification up to the temperatures where other sintering process become effective.

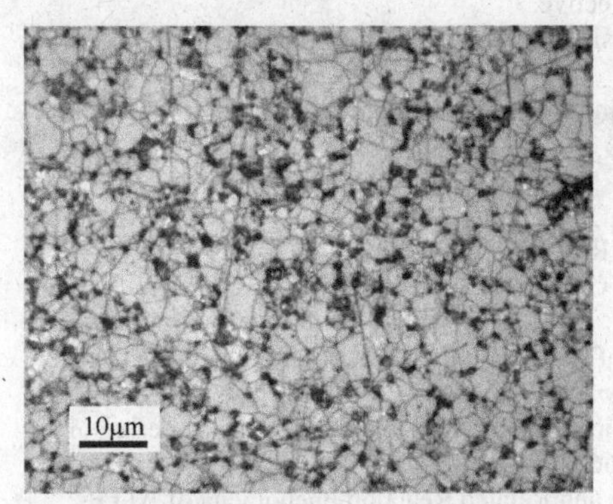

Figure 6. Optical micrograph of an etched surface of the B_4C ceramics sintered at 2104°C in an Ar atmosphere containing gaseous Al and Si species.

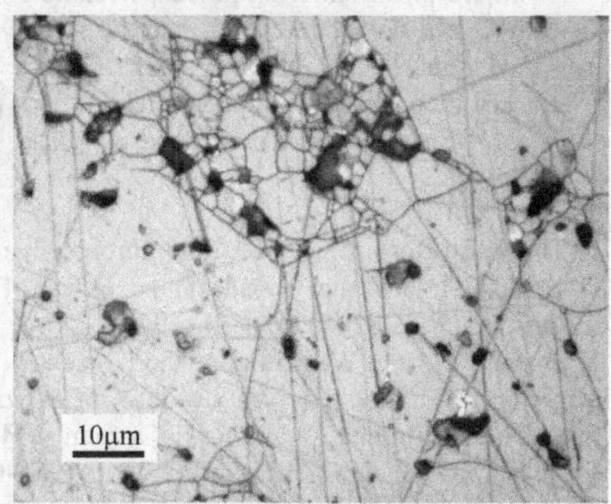

Figure 7. Optical micrograph of an etched surface of the B_4C ceramics sintered at 2187°C in an Ar atmosphere containing gaseous Al and Si species.

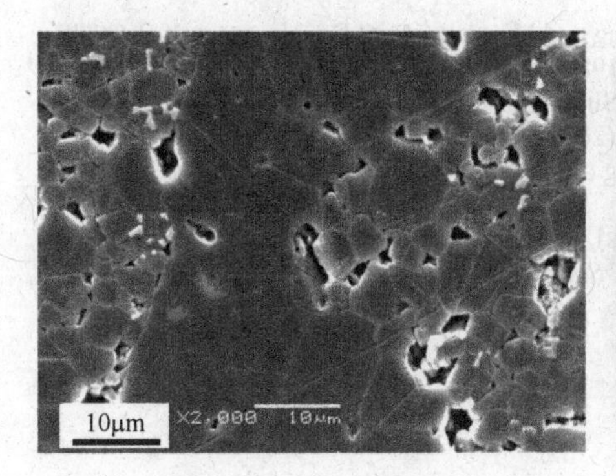

Figure 8 (a). SEM micrograph of an etched surface of the B_4C ceramics sintered at 2187°C in an Ar atmosphere containing gaseous Al and Si species.

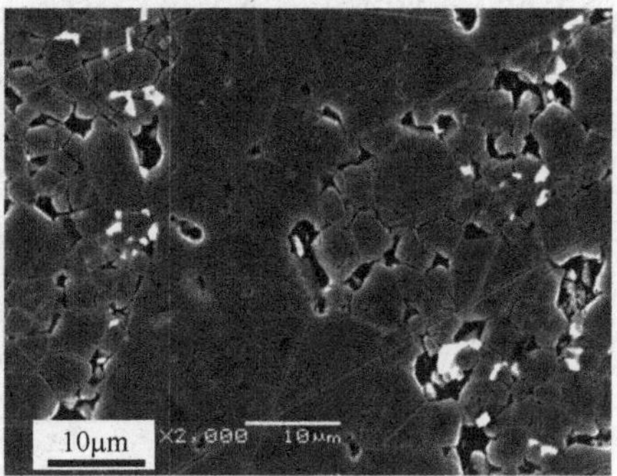

Figure 8 (b). Back-scattered electron micrograph of an etched surface of the B_4C ceramics sintered at 2187°C in an Ar atmosphere containing gaseous Al and Si species.

CONCLUSION

Boron carbide (B_4C) was pressurelessly sintered at temperatures ranging from 2104 °C to 2226°C in an Ar atmosphere containing gaseous Al and Si species. The crystalline phases in the densified samples were analyzed with XRD, followed by the microstructural observations with optical microscopy and SEM. The following results were obtained.

1. Gaseous Al and Si species infiltrated into the B_4C compact during the sintering. Thin B_2O_3 films on the B_4C particle were removed by the reduction with gaseous Al species, which hindered neck growth by the vapor-phase transport with an assist of B_2O_3. As a result, the driving force for densification was preserved up to the temperatures where liquid-phase sintering process became effective.
2. SiC and Al_4SiC_4 were formed during the sintering as the results of the reaction of B_4C with gaseous Al and Si species.
3. The mechanism of the densification was inferred to be the liquid-phase sintering because of the irregular-shaped secondary phase.

REFERENCES

[1]F. Thevenot, Boron Carbide- A Comprehensive Review, *J. Eur.Ceram.*, **6**, 205-25 (1990).
[2]H. Suzuki, H. Hase and T. Maruyama, Effect of Carbon on Sintering of Boron Carbide, *J. Ceram. Soc. Jpn.*, **87** (8), 430-33 (1979).
[3]K. A. Schwetz and W. Grellner, The Influence of Carbon on the Microstructure and Mechanical Properties of Sintered Boron Carbide, *J. Less-Common Met.*, **82**, 37-47 (1981).
[4]S. L. Dole, S. Prochazka and R. H. Doremus, Microstructural Coarsening During Sintering of Boron Carbide, *J. Am. Ceram. Soc.*, **72** (6), 958-66 (1989).
[5]C. H. Lee and C. H. Kim, Pressureless Sintering and Related Reaction Phenomena of Al_2O_3-Doped B_4C, *J. Mater. Sci.*, **27**, 6335-40 (1992).
[6]H-W. Kim, Y-H. Koa and H. E. Kim, Densification and Mechanical Properties of B_4C with Al_2O_3 as Sintering Aid, *J. Am. Ceram. Soc.*, **83** (11), 2863-65 (2000).
[7]L. S. Sigl, Processing and Mechanical Properties of Boron Carbide Sintered with TiC, *J. Eur.Ceram.*, **18**, 1521-29 (1998).
[8]D. K. Kim and C. H. Kim, Pressureless Sintering and Microstructural Development of B_4C-TiB_2 Composites, *Adv. Ceram. Mater.*, **3** (1), 52-55 (1988).
[9]H. Lee, R. F. Speyer and W. S. Hackenberger, Sintering of Boron Carbide Heat-Treated with Hydrogen, *J. Am. Ceram. Soc.*, **85** (8), 2131-33 (2002).
[10]H. Lee and R. F. Speyer, Pressureless Sintering of Boron Carbide, *J. Am. Ceram. Soc.*, **86** (9), 1468-73 (2003).
[11]T. Kumazawa, T. Honda, Y. Zhou, H. Miyazaki, H. Hyuga and Y. Yoshizawa, Pressureless Sintering of Boron Carbide Ceramics, *J. Ceram. Soc. Jpn.*, **116** (12), 1319-21 (2008).

PROCESSING STRATEGY FOR PRODUCING ULTRA-HIGHLY POROUS CORDIERITE

Manabu FUKUSHIMA, Masayuki NAKATA and Yu-ichi YOSHIZAWA

National Institute of Advanced Industrial Science and Technology (AIST)
2266 Anagahora, Shimo-Shidami moriyama-ku, Nagoya, Aichi 463-8560 Japan

ABSTRACT

Cordierite with ultrahigh porosity and unidirectionally oriented cylindrical cells was prepared using a gelation freezing method (GF method). Cordierite powder was mixed with a gelation agent, water (pore former) and the resulting gel was frozen at -20 or -50 ℃, dried using a freeze drier under vacuum, then degreased at 600℃ and sintered at 1200-1400℃ for 2 hr. The porosity was determined to be 59-93%, and was dependent on both the solid load in the slurry and the sintering temperature. Scanning electron microscopy revealed a microstructure of unidirectionally oriented micrometer-sized cylindrical cells in the sintered body. The cell size was 25-270μm, which were depending on the freezing temperature. The numbers of cells in the cross sections of the sintered bodies (frozen at -20 and -50℃) were 15 and 1500cells/mm^2, respectively. The resulting cordierite had a total porosity of 87%, compressive strength of 4.1MPa, and 207% water adsorption, which was attributed to tight packing of particles and good pore connectivity. The porosity, number of cells, strength and water adsorption were significantly higher than that values for other reported porous cordierite ceramics.

INTRODUCTION

The incorporation of organic pore formers such as PMMA (polymethylmethacrylate) and carbon beads into raw ceramic powder, and their subsequent removal by oxidation, is known to be effective for improvement of porosity in porous ceramics. However, this method accompanies the evolution of large amounts of carbon oxide gas species and a time-consuming to prevent cracking of the ceramic, and also control of the pore size and its orientation is difficult due to the size of pore former. Therefore, a fabrication process for ultra-highly porous ceramics that enables control of the porosity, pore size and its orientation, and is eco-friendly (short processing periods and pore formers without organic components) is desirable.

Many applications are expected for highly porous ceramics, such as filters, catalyst supports, thermal insulators, adsorbents, shock absorbers and lightweight components. Among oxide ceramics, cordierite has attracted much attention due to its low thermal expansion coefficient, excellent thermal shock, and chemical stability at elevated temperatures, in addition to its low cost. Some porous cordierite has been utilized as filter and catalyst support in industry. Therefore, a process for the fabrication of ultra-highly porous cordierite with oriented pores and controlled pore size is highly desirable.

In order to fabricate ultra-highly porous cordierite, we have focused on the gelation-freezing (GF) route, which has advantages of porosity and pore orientation control [1-3]. In this method, pores are formed by ice due to the freezing of water in the slurry, which makes this an environmentally friendly process compared to other methods that require the addition of organic pore formers. First, raw ceramic powder is dispersed to form a gelation agent. After gelation, the wet gel, which retains water, is frozen, and formation of ice crystals is observed. A vacuum freeze drier is used to dry the frozen gel, so that ice is sublimated and removed without shrinkage. Finally, the green body is sintered and a porous ceramic is obtained. This process does not require large amount of organic components. In addition, control of the porosity is expected by varying the solid load in the slurry, and pore orientation may be tailored by directional freezing.

The fabrication of porous ceramics by the freezing route has been reported using a water-based slurry or preceramic polymer gel as the starting chemical. Nishihara et al. reported the preparation of silica gel with a surface area of 900 m^2/g using silane [4-5]. Ding et al. reported the preparation of mullite with porosity of around 90% from an alkoxide precursor [6]. On the other hand, Fukasawa et al. reported a water based slurry route for the production of alumina with 40-80% porosity and silicon nitride with 50-70% porosity [7-10]. However, high porosity was not available for the water-based slurry route, due to the difficulty in handling the green body and degreased body, whereas for the polymeric precursor route, the material obtained was limited by the type of precursor used. These are problems to overcome for successful utilization of the freeze-casting process.

The GF route that we have previously reported [1-2] may be a solution to these problems. Handling of the green body was facile due to polyethylenimine (PEI) in the gel, and porous material could be prepared regardless of the type of starting powder. However, the resultant porosity was at most 70%, due to the water retention ability of the PEI gel. Therefore, the use of gel that retains a large amount of water is proposed for the preparation of highly porous ceramics. In the present work, the preparation of ultra-highly porous cordierite with oriented cylindrical pores and controlled pore size was attempted by the GF method, and preliminary results for open porosity, pore size and microstructure are presented.

EXPERIMENTAL PROCEDURE

Commercially available cordierite powder with an average particle size of 1.7μm was used as a starting powder (Marusu Glaze Co., Ltd., Aichi, Japan). The cordierite powder was poured into a gelatin solution (Wako Pure Chemical Industries Ltd., Tokyo, Japan) at around 50℃. The mixing ratios of raw powder/gelatin solution were 5/95 and 20/80 by volume, and these solutions are referred to as C95 and C80, respectively. The slurry was mixed using a planetary homogenizer (AR-250, Thinky, Tokyo, Japan), and then the slurry was cast into a plastic mold and cooled at 7℃ for 10 h, (which time gelation of the slurry was observed). The mold was then immersed into an ethanol bath at -50℃ or -20℃. After removal of the mold, the frozen gel containing was dried by a vacuum freeze drier at 10-35℃ for 24hr (FDU-2100, Tokyo Rikakikai Co., Ltd., Tokyo, Japan) to obtain ceramic green bodies. The green bodies were degreased at 600℃ for 2hr and sintered at 1200-1400℃ for 2 h using a heating rate of 5℃/min.

The apparent densities of the sintered specimens were measured by water displacement with the Archimedes method, according to equation (1),

$$d = \frac{\rho W_a}{W_w - W_l} \qquad (1)$$

where Wl, Ww and Wa indicate the weight in water, wet weight and dry weight, respectively, and ρ is the water density at the measurement temperature. The microstructures of fractured or polished surfaces were observed by scanning electron microscopy (SEM; JSM-5600, JEOL, Tokyo, Japan). The cell size and number of cells were measured using the intercept method or image analysis of SEM micrographs, and approximately 400micrographs and more than 11000cells were investigated for characterization. Pore size distributions were measured by mercury intrusion porosimetry (PoreMaster-GT, Yuasa Ionics Inc., Osaka, Japan), which measures the pore size in the range of 6.1nm-426μm. X-ray computed tomography (CT; SMX-130CT-SV, Shimadzu, Kyoto, Japan) was measured with operation at

68kV and 68A. Three-dimensional (3D) images and pore density of each image were analyzed using modeling software (VG-STUDIO, NVS, Tokyo, Japan).

RESULTS AND DISCUSSIONS

Figure 1 shows the porosity of cordierite (frozen at -50°C) derived from different solution content in the slurry. Specimen C95 exhibited 93-87% porosity and specimen C80 exhibited 76-59% porosity for sintering at 1200-1400°C. The porosity was affected by the sintering temperature and slurry concentration. The porosity decreased with increasing sintering temperature, which was similar to the densification for the normal sintering route. When sintered at 1200°C, the porosity was almost consistent with the solution content (95 and 80%) in the slurry, due to low shrinkage at this sintering temperature. On the other hand, closed porosity for all samples was less than 1.0%, which suggests that during drying, the ice is almost completely converted to pores.

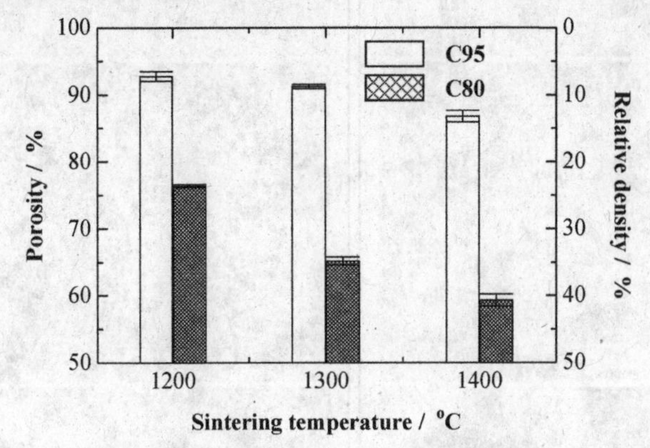

Figure 1. Porosity of cordierite derived from different solution content in the slurry and sintered at 1200-1400°C.

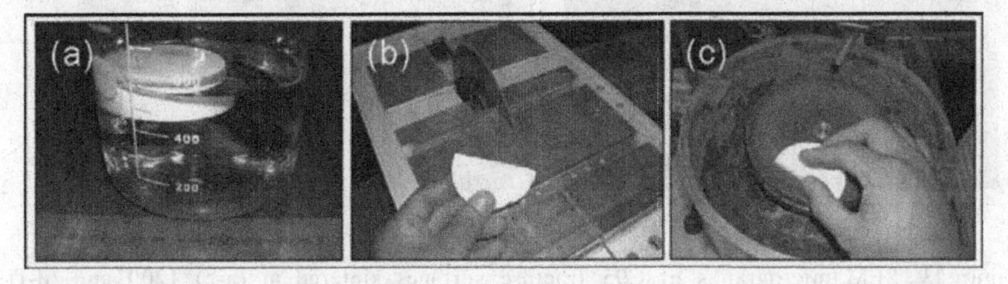

Figure 2. Photographs of (a) cordierite with 93% porosity floating in water, and the (b) cutting and (c) polishing of the same cordierite.

Figure 2 shows photographs of 93% porous cordierite, and cutting and polishing of the same material. A 70 mm diameter disk of the cordierite could be floated on water due to ultra-highly porosity. The shape of porous sample was determined by the shape of the mold used, as for casting methods. It is particularly noteworthy that cutting and polishing are available for the porous specimens.

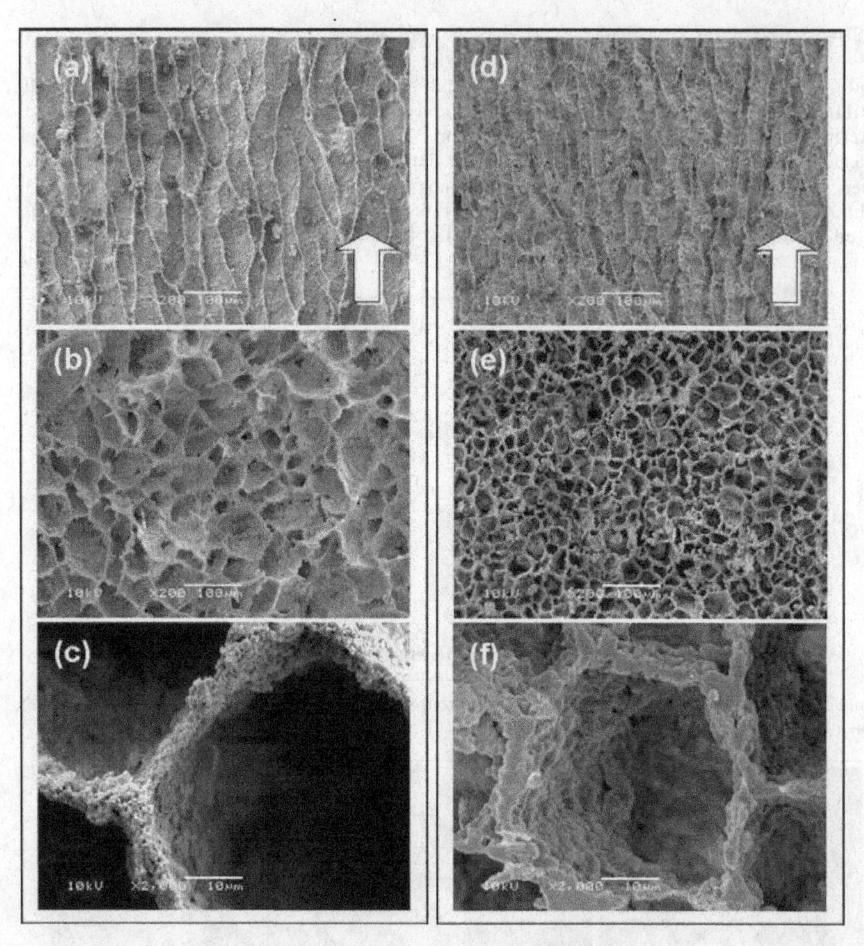

Figure 3. SEM micrographs of C95 fracture surfaces sintered at (a-c) 1200 and (d-f) 1400 ℃; parallel to (a and d)the direction of freezing (-50 ℃), as indicated by the arrows, and perpendicular to (b-c and e-f) the direction of freezing (-50 ℃).

Figure 3 shows typical SEM micrographs of fracture surfaces of the C95 sample sintered at 1200 and 1400℃; in parallel and perpendicular to the direction of freezing (-50 ℃). Highly oriented cylindrical cells were found to form parallel to the direction of freezing. These cells are thought to be formed due to the growth of ice crystals. During freezing, the ceramic particles were observed to be unidirectionally pushed alongside the direction of the growing ice crystal. This structure was present in the entire sintered body, and the formation of dendritic structures due to ice crystal growth was not observed, but is well known in the freeze casting of water-based slurry (without organic components) [6-10]. For the perpendicular direction, cylindrical-shaped cell (b and e) were observed. The cellular honeycomb structure consisted of micrometer-sized cells, of which the shape was different to that of foam, pore-forming agents and those observed in other freeze-casting routes [6-10, 11-13]. On the other hand, at high magnification (Fig. 3(f)), dense cell walls were observed for the specimen sintered at 1400℃.

The average cell diameters of the C95 specimens sintered at 1200 and 1400 ℃ were 46 and 25μm, respectively, as calculated by image analysis. The pore size decreased with increasing sintering temperature, due to shrinkage during sintering. The number of cells in the cross section was approximately 500 and 1500cells/mm^2 for C95 sintered at 1200 and 1400℃, respectively, which increased with decreasing cell size. In addition, the thickness of the cell wall was 4.4 and 3.0μm for the specimens sintered at 1200 and 1400℃. The cell wall thickness decreased with increased sintering temperature, which was found to be due to the densification of grains during sintering.

Figure 4 shows a typical example of pore size distribution of the C95 specimens sintered at 1200 and 1400℃ (frozen at -50 ℃). For the C95 specimen sintered at 1200 ℃, the pore size distribution was bimodal with peaks around 0.6 and 20μm. The smaller pores are attributed to the spaces between particles in the cell wall, whereas the larger cells are those formed by ice freezing. The cell size was smaller than that observed by SEM, which indicates mercury intrusion into the bottle-necks of some pores, which is the well-known "ink-bottle" effect. In contrast, the distribution of the C95 specimens sintered at 1400℃ showed only one peak at approximately 16μm, which indicates the densification of particles in the cell wall, as shown in the SEM images of Fig. 3. The cell size (attributed to the formation of ice crystal) was found to decrease with increasing sintering temperature, which is also consistent with the SEM observations from Fig. 3.

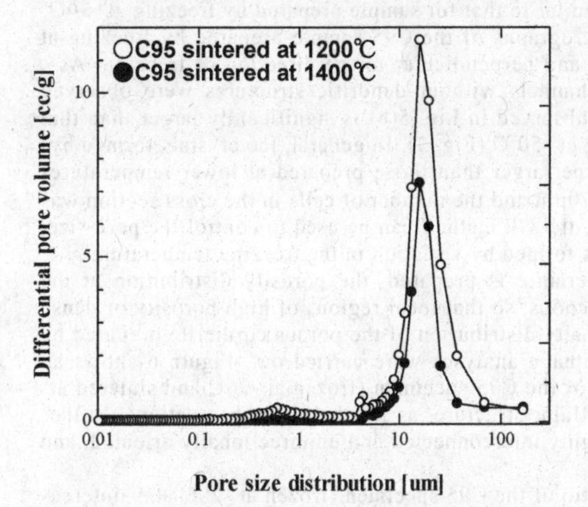

Figure 4. Typical pore size distributions for C95 specimens sintered at 1200 and 1400 ℃ (frozen at -50 ℃).

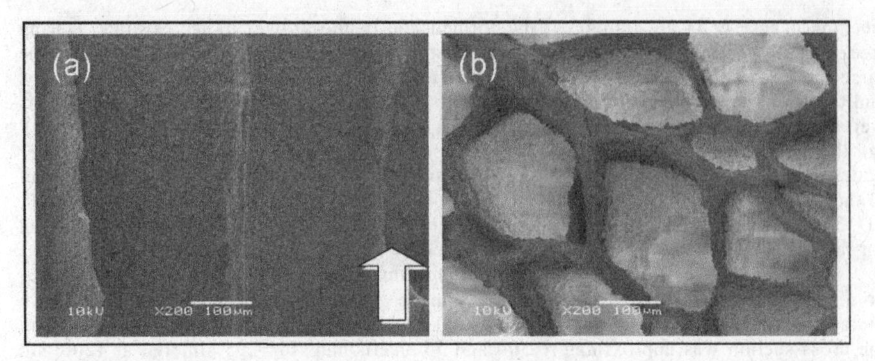

Figure 5. SEM micrographs of the C95 sample (frozen at -20℃) sintered at 1400℃; (a) parallel to the direction of freezing, as indicated by the arrows, and (b) perpendicular to the direction of freezing.

In order to investigate the effect of freezing temperature on cell size, specimens frozen at -20℃ and sintered at 1400℃ were prepared. The resulting porosity was approximately 86%, which was very similar to that for sample prepared by freezing at -50℃ (see Fig. 1). Figure 5 shows SEM micrographs of the C95 sample prepared by freezing at -20℃ and sintering at 1400℃; parallel and perpendicular to the direction of freezing. As shown in Fig. 5(a), interconnected channels without dendritic structures were observed throughout the samples. The cell size observed in Fig. 5(b) is significantly larger than that for the specimens prepared by freezing at -50 ℃ (Fig. 3). In general, ice crystals formed by freezing at high temperature seem to be larger than those prepared at lower temperatures [14]. The average pore diameter was 270μm and the number of cells in the cross section was approximately 15cells/mm^2. Therefore, the GF method can be used to control the pore size, due to the different sizes of ice crystals formed by variation in the freezing temperature.

When a ultra-highly porous ceramic is prepared, the porosity distribution at the micrometer level is not usually homogenous, so that local regions of high porosity or dense regions are present. To clarify the porosity distribution of the porous cordierite prepared by the GF method, X-ray CT scans and image analyses were carried out. Figure 6 shows an X-ray CT micrograph and a 3D image of the C95 specimen (frozen at -20℃ and sintered at 1400℃). The micrograph shows the cellular structure, as for the SEM observations. In the 3D image, the cells are shown to be highly interconnected and unidirectionally oriented, and with uniform size.

Figure 7 shows the cell area ratio of the C95 specimen (frozen at -20℃ and sintered at 1400℃) plotted from the bottom to the top of the observed part (shown in Fig. 6(a). All micrographs (ca. 450 micrographs) were converted to black-and-white images and the cell/wall area ratios were calculated for each micrograph and plotted. The results showed a cell/wall area ratio of approximately 84% and almost constant values were obtained in the measurement part (ϕ2.8mm × 2.5mm), which suggests that the porosity at the micrometer level was constant. Therefore, the GF method can provide a porous body with constant porosity and uniform porosity distribution at the micrometer level.

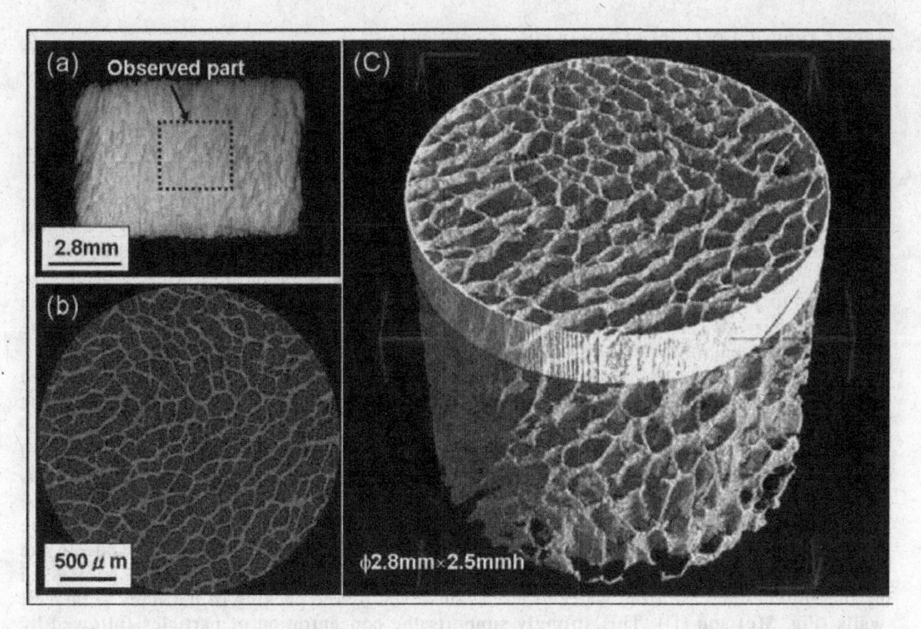

Figure 6 (a)photograph of the C95, (b)micrograph and (c) 3D image (frozen at -20℃ and sintered at 1400℃).

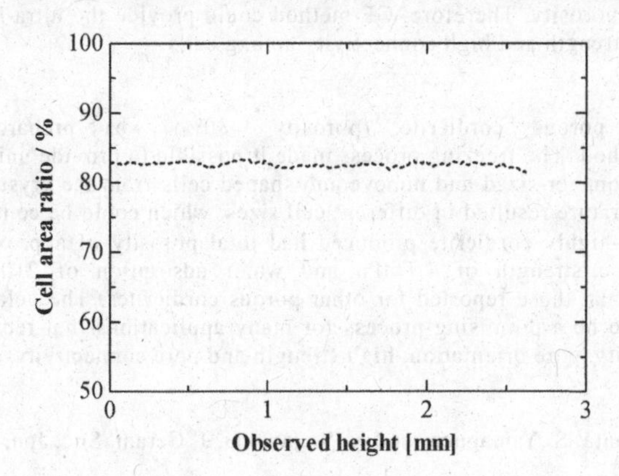

Figure 7. Cell/wall area ratio obtained from X-ray CT micrographs of the C95 specimen frozen at -20℃ and sintered at 1400℃. Approximately 450 micrographs from the bottom to the top of the observed part were used.

Table1 Typical properties of highly porous cordierite.

	This study	Song	Senguttuvan	Kato
Porosity (%)	86.8	~74	83	80
Cell size (um)	25	13	20-40	21
Compressive strength (MPa)	4.1	~4.3	0.6	-
Water absorption	207%	-	-	100%

Table 1 summarizes the typical properties of the highly porous cordierite obtained in this study, including reported results from three other studies [15-17]. (Note that the report by Kato et al. [17] is for aluminosilicate.) Song et al. [15] reported cordierite with a porosity of 74%, pore size of approximately 13μm and compressive strength of 4.3MPa; the cell size is similar to that obtained in the present work. Senguttuvan et al. [16] reported cordierite with a porosity of 83%, cell size around 20-40μm and compressive strength of 0.6MPa. The compressive strength of the C95 sample (frozen at -50℃ and sintered at 1400℃) with a total porosity of 86.8% and a cell size of 25μm was 4.1MPa, which was higher than that obtained by other reported methods, regardless of the higher porosity. This is thought to be due to the high packing density of the raw powder.

Previous research on freeze-casting suggested that the raw powder is highly packed upon freezing [13]. Araki and Halloran [13] showed that porous alumina (41% porosity) was prepared by freezing a slurry with camphene, and a dendritic microstructure by the solidification of camphene was observed. Figure 3 shows that particles were pushed aside along the direction of freezing. During freezing, the particles are concentrated due to the removal of water, that is, by phase separation. The particles are then thought to be pressed by the volume increase from water to ice, as observed by the packing of particles in the cell walls (Fig. 3(c) and (f)). This strongly supports the concentration of particles followed by packing. In addition, water adsorption was significantly larger than that of the previous report, which is closely correlated with high fluid permeability and good pore connectivity as well as ultra-high porosity. Therefore, GF method could provide the ultra-highly porous cordierite with high strength and high connectivity among cells.

CONCLUSIONS

Ultra-highly porous cordierite (porosity >80%) was prepared using a gelation-freezing method. The freezing process made it possible to provide unidirectionally, highly oriented, micrometer-sized and honeycomb-shaped cells from ice crystals. Variation of the freezing temperature resulted in different cell sizes, which could be controlled within 25-270μm. The ultra-highly cordierite produced had total porosity of approximately 87% and high compressive strength of 4.1MPa and water adsorption of 210%, which is significantly higher than those reported for other porous cordierites. The gelation freezing method is expected to be a promising process for many applications that require ceramics with ultra-high porosity, pore orientation, high strength and pore connectivity.

REFERENCES

[1] M. Nakata, K. Tanihata, S. Yamaguchi and K. Suganuma, J. Ceram. Soc. Jpn, 113, 712-715 (2005).
[2] M. Nakata, M. Fukushima and Y.Yoshizawa, Ceram.Eng.Sci.Proc., 28, 7, 139-144 (2008).
[3] M. Fukushima, M. Nakata and Y.Yoshizawa, J. Ceram. Soc. Jpn, 116, 1322-1325 (2008) .
[4] S. R. Mukai, H. Nishihara and Tamon, Microporous and Mesoporous Mater., 63, 43-51 (2003).
[5] H. Nishihara, S. R. Mukai, D. Yamashita and H. Tamon, Chem. Mater., 17, 683-689 (2005).

[6] S. Ding, Y.-P. Zeng and D. Jiang, J. Am. Ceram. Soc., 90, 2276-2279 (2007).
[7] T. Fukasawa, Z.-Y. Deng, M.Ando and T.Ohji, J. Ceram. Soc. Jpn, 109, 1035-1038 (2001).
[8] T. Fukasawa, M.Ando and T.Ohji, J. Ceram. Soc. Jpn, 110, 627-631 (2002).
[9] T. Fukasawa, M.Ando, T.Ohji and S. Kanzaki, J. Am. Ceram. Soc., 84, 230-232 (2001).
[10] T. Fukasawa, Z.-Y. Deng, M.Ando, T.Ohji and S. Kanzaki, J. Am. Ceram. Soc., 85, 2151-2155 (2002).
[11] H. Schmidt, D. Koch, G. Grathwohl and P. Colombo, J. Am. Ceram. Soc., 84, 2252-2255 (2001).
[12] S. Bhattacharjee, L. Besra and B. P. Singh, J. Eur. Ceram. Soc., 27, 47-52 (2007).
[13] K. Araki and and John W. Halloran, J. Am. Ceram. Soc., 88, 1108-1114 (2005).
[14] S. Ueno, R. Shirakashi, G. Do, Y. Sagara, K. Kudoh and T. Higuchi, Trans. of the JSRAE, 21, 337-344 (2004). (in Japanese)
[15] I.H.Song, M.J.Kim, H.D.Kim and Y.W.Kim, Scripta Mater., 54, (2006) 1521–1525.
[16] T.D. Senguttuvan, H.S. Kalsi, S.K. Sharda and B.K. Das, Mater.Chem.Phys., 67, (2001) 146-150.
[17] T.Kato, K.Ohashi, M.Fuji and M.Takahashi, J. Ceram. Soc. Japan, 116, (2008), 212-215.

ISSUES IN THE SYNTHESIS AND FABRICATION OF REFRACTORY CARBIDES, BORIDES, SILICIDES AND THEIR MIXTURES

A.K. Suri, N. Krishnamurthy and C. Subramanian,
Materials Group, Bhabha Atomic Research Centre,
Mumbai- 400085, Maharashtra, India

ABSRATCT

Among the non-oxide ceramics, certain carbides and borides possess attractive combination of properties and good potential for applications but are difficult to fabricate, because of their physico-chemical and mechanical characteristics. The synthesis of boron carbide, borides of titanium, zirconium, hafnium, chromium and rare earth elements, their fabrication to consolidated pieces and determination of their properties in various stages of processing as well as in the final form, have been investigated. While the selection of the compounds for investigation was guided by their immediate technological applications, working with them brought to the fore several issues relating to their preparation and processing and these are covered in this paper. Boron carbide as well as the borides were synthesized from constituent elements, by chemical reaction of compound intermediates and also by fused salt electrolysis. Fine powder was obtained from the as reduced material by mechanical milling using special equipments. The densification by cold compaction and sintering as well as by hot pressing was investigated. Property measurements on the final products were carried out to relate the present material with the standards. The possibility of synthesizing and mixing silicides of refractory metals with borides for enhancing the high temperature oxidation resistance of borides was also investigated.

INTRODUCTION

Boron carbide and refractory metal borides have attractive combination of properties such as low density, high melting point and hardness, chemical inertness, neutron absorption ability and excellent thermal and electrical characteristics making them potential materials for many advanced applications. Selected properties are listed in table 1.[1-6]

Table.1 Properties of refractory materials[1-6]

Material	Density g/cc	Melting point °C	Hardness GPa	Young's modulus GPa	Fracture toughness, MPam$^{1/2}$	Flexural strength MPa	Thermal conductivity W/m/K	Electrical resistivity μΩ-cm	Coefficient of thermal expansion 10^6 /K
B_4C	2.52	2450	28-37	450-470	3.0 – 3.5	300	30 - 40	10^6	5.0
TiB_2	4.52	3225	25-35	560	5.0 - 7.0	700-1000	60 - 120	10-30	7.3
ZrB_2	6.1	3245	22-26	300-350	4.0 – 6.0	300	20 - 58	9.2	6.8
HfB_2	11.21	3380	21-28	500	-	350	104	11	6.3
EuB_6	4.99	2580	18-26	-	-	183	23	85	6.9
CrB_2	5.20	2200	11-20	211	-	600	32	30	10.5
$MoSi_2$	6.30	2050	13	384	2.0-2.5	-	50-221	21	8.4

Apart from well established uses as abrasive and wear resistant material, neutron absorber, personnel and vehicle armor material, the possible use of boron carbide in high temperature electronic devices and also for a Be/Be alloy replacement in aerospace applications are considered.[7-9] Composite material containing boron carbide can become relevant for the first wall of nuclear fusion reactors.[10-13] TiB_2 is a potential candidate material for high-temperature structural applications and as control rod material for high temperature nuclear reactors.[2,4] ZrB_2 and HfB_2 are suitable materials for service in the extreme thermal and chemical environments associated with hypersonic flight, atmospheric re-entry vehicles and rocket propulsion.[14-16] Chromium diboride is a valuable additive to borides used at high temperatures for improving their oxidation resistance and an excellent coating material where resistance to both wear and corrosion is required.[17] Hexaborides of rare earth elements such as LaB_6, EuB_6, SmB_6, YB_6 etc, are the materials highly suitable for neutron absorber, fluorine rich environments and also an effective additive for improving the oxidation resistance of refractory metal borides.[18,19] $MoSi_2$ is an excellent structural and electro conductive ceramic that fulfills many of the conditions for high temperature applications.[20] Use of $MoSi_2$ as a sinter additive improves the sinterability and fracture toughness of the composite borides.

This paper covers the process of synthesis and consolidation of the above materials as developed and practiced in our laboratory. Boron carbide and molydisilicide were synthesized from the elements and the borides of titanium, zirconium, hafnium, chromium and rare earth elements were prepared by the reaction of their respective oxides with boron carbide. Syntheses of compounds were carried out in a vacuum induction furnace. The initial experimental conditions were based on thermogravimetric studies. As most of the reactions were based on solid - solid contacts, the reactants were fine ground and mixed thoroughly and compacted. The compacts were placed in a graphite crucible and subjected to progressive heat soaking. Carbides and borides formed were crushed to fine powder for densification either by pressureless sintering or by hot pressing. For pressureless sintering the green compacts were heated in a vacuum induction furnace for a specified period and at fixed temperature. Hot pressing of the powder was carried out in a vacuum hot press using high density graphite die and plunger. The progress of densification was monitored by LVDT (Linear Variable Differential Transducer). The temperature was measured using a two color pyrometer. The product was characterized by XRD, density measurement and hardness. Mechanical, thermal and electrical properties of pellets with >95% ρ_{th} were evaluated. All the operations were carried out initially in batches of ~100 g and selected experiments with larger batch sizes.

BORON CARBIDE

Synthesis

Boron carbide is a compositionally disordered material that exists as rhombohedral phase in a wide range of composition, which extends from $B_{10.4}C$ [8.8 atom % C] to B_4C [20 atom % C].[21] Among them, B_4C is superior in properties such as hardness and thermal conductivity. Since B_4C is in equilibrium with free carbon, synthesis of B_4C without free carbon is a challenge.[22] The analytical study of B-C system also is difficult due to extreme hardness and chemical stability of boron and boron carbide phases.[23] Boron carbide is commercially produced by carbothermic reduction of boric acid in electric arc or graphite resistance furnaces.[24] In these furnaces the temperature of reaction is high (>2000°C) and yield of boron carbide is poor due to evaporation losses. The product quality also is not uniform, varying in carbon content due to non uniform temperature in the reaction zone. Boron carbide can also be produced by magnesiothermic reduction of boron oxide in presence of carbon,[25,26] but the magnesium content of the carbide is high due to the formation of magnesium boride during the reduction.

For use as control rod material in fast breeder reactors it is necessary to have B^{10} enriched B_4C, which has high absorption cross section for fast neutrons. Carbothermic process is not viable due to boron losses which is unacceptable when B^{10} is involved. The synthesis of B_4C from pure enriched boron and carbon is practiced.[27,28] During synthesis from the elements, it is possible to control the composition, carbon content and phase purity. Densification of boron carbide is difficult due to the predominant covalent bonding in the compound. Mass transport mechanisms such as grain boundary and volume diffusion that aid densification are effective only at temperatures above 2000°C at normal pressures.[9] However, hot pressing helps in achieving higher densification at lower temperatures.[9]

Experiments on the synthesis of boron carbide from its elements were carried out by heating compacted mixtures between 1200 to 1950°C at 0.01 Pa for different durations. The product obtained upto 1600°C was completely amorphous and at above 1750°C crystalline boron carbide was obtained. The XRD peaks of boron carbide phase (rhombohedral) were weak even at 1750°C, but dominant at higher temperatures. Although formation of boron carbide from its elements has a large negative free energy change, the heat of reaction (ΔH^{o}_{298} = -39 kJ/mol) is not sufficient for a self-sustaining reaction. Moreover, the generally slow kinetics of solid–solid reaction is improved by a rise in temperature. Synthesis of boron carbide with varying B/C ratio (4.0 to 4.6) in the charge was therefore carried out at 1850°C for 2h. At a B/C ratio of 4.0 the product contains B_4C and graphite. With higher boron content in the charge, the product still contained graphite, but the excess boron helped in the formation of boron rich B_8C. Conditions for synthesis of boron carbide and refractory borides are given in table 2. Presently, all the boron carbide produced in our laboratory for control rod application contains a small amount of free carbon.

Table 2 conditions for synthesis of boron carbide and refractory borides

Sr.no	Compound synthesized	Charge composition	Temperature °C	Phases obtained
1	B_4C	B+C (B/C = 4)	≤1600	Amorphous
			1800	B_4C + Graphite
		B+C (B/C > 4)	1800	$B_4C + B_8C$ + Graphite
2	TiB_2	$TiO_2 + B_4C$ +C	1230	Ti_3B_4, TiO_2, Ti_2O_3 + C
			1500	TiB_2 Impure
			≥ 1800	TiB_2 Pure
3	ZrB_2	$ZrO_2 + B_4C$ +C	1800	ZrB_2, ZrB + C
		modified charge ratio	1800	ZrB_2
4.	HfB_2	$HfO_2 + B_4C$ +C	1825	HfB_2
5	CrB_2	$Cr_2O_3 + B_4C$ +C	1700	CrB_2, Cr_3B_4 + C
		modified charge ratio	1700	CrB_2
6	$MoSi_2$	Mo + Si	1400	$MoSi_2$
7	EuB_6	$Eu_2O_3 + 3B_4C$	1400	EuB_6
8	YB_6	$Y_2O_3 + 3B_4C$	1700	$YB_6 + YB_4$

DENSIFICATION

Pressureless sintering

Densification of boron carbide powder was studied in our laboratory by pressureless sintering with and with out additives and also by hot pressing.[29,30] For pressureless sintering, commercial boron carbide and for hot pressing, B_4C synthesized from elements were used. Particle size of boron carbide powder and the sintering temperature are two key parameters in densification. While pressureless sintering of B_4C powder compacts can be done under an inert gas cover, the use of vacuum appears to have beneficial effects. By pressureless sintering, upto 2300°C a maximum density of 85%ρ_{th} was obtained when the carbide starting particle size was in the range of 0.5–2.0 μm. Compact densities of above 90%ρ_{th} could be achieved by sintering at 2375°C using finer particles of median diameter 0.5/ 0.8 μm.

Free carbon in the carbide is useful in reducing the oxide layer on the boron carbide powder, thereby promoting sintering and suppressing grain growth.[31,32] Carbon can be introduced in the form of an additive such as phenol-formaldehyde resin, which serves as binder in cold pressing and as a precursor for uniformly distributed carbon in the compact.[31] Addition of carbon helped in achieving densities of 90–91%ρ_{th} at 2325°C. It appears to act as grain refiner also. ZrO_2 as a sintering additive was found to be very effective in reducing the sintering temperature.[30] Samples sintered at 2275°C for 60 minutes with the addition of 5 wt.% ZrO_2 showed densities in the range of 93–96%ρ_{th}, compared to 86.6%ρ_{th} for compacts without the additive. Hardness of the samples sintered with ZrO_2 additive was higher at 30–31.5 GPa, compared to 27 GPa of B_4C obtained without additive. Both carbon and zirconium are neutron transparent, hence could be used with boron carbide for nuclear applications. Conditions for densification of boron carbide and refractory borides are presented in table 3.

Hot pressing

Particles of median diameter 1-3 μm was hot pressed at 1900°C and 35MPa pressure to give a compact of density of >95%ρ_{th}. The properties of a typical boron carbide sample prepared in our laboratory are: Density – 2.4 g/cc, Hardness - 28 GPa, Fracture toughness (indentation) - 4.1 $MPa.m^{1/2}$, Young's Modulus - 420 GPa, Flexural Strength at RT - 363 MPa, Thermal conductivity at 50°C - 28.78 WM^1K^{-1}, Electrical Resistivity at RT -1.27 Ω-cm. These properties are matching with the literature values for similar use.

TITANIUM DIBORIDE

Synthesis

Synthesis of titanium diboride was carried out in our laboratory by boron carbide reduction of titanium dioxide.[33] The free energy of the overall reaction

$$2TiO_2+B_4C+3C= 2TiB_2 + 4CO \quad (1)$$

is negative above 1257K.[33] The use of vacuum makes the reaction feasible at lower temperatures. Experiments were conducted at temperatures between 1200 to 1800°C in 0.04 Pa. Even though at 1230°C itself TiB_2 formation was observed, the product was a mixture of Ti_3B_4, TiO_2, Ti_2O_3 and C. After reaction at 1360°C, TiO_2 phase was no more present and at 1500°C and above the reaction is nearly complete, but the product contained carbon and oxygen as impurity (each 2-3%). Further removal of these impurities need a purification by heating to a temperature of >1800°C and a higher vacuum of 0.01 Pa. Even extended hours of soaking at lower temperatures do not yield a pure product. Only above 1800°C and 0.01 Pa, TiB_2 obtained had less carbon (0.6%) and oxygen (0.5%).

Consolidation

The applications of TiB_2 are probably limited by the challenges in sintering the material to the required density.[2] Low diffusivity, grain growth at high temperature and presence of oxide layer on the surface of the particles are some reasons for poor sinterability. Without an additive, TiB_2 could be densified only at high temperatures by hot pressing. Densification of TiB_2 powders by pressureless sintering in a reducing atmosphere was attempted upto 1900°C but the density attained was only 91%ρ_{th}. By hot pressing, pellets of 88 and 97.6%ρ_{th} density were obtained at 1700 and 1800°C respectively. Hardness and fracture toughness values of high density sample were measured as 26 GPa and 5.3 MPa.m$^{1/2}$. Thermal conductivity was measured to be 62.5 W/m/K at RT. The property values are in good agreement with the reported literature values. $MoSi_2$ with a comparatively low melting point and good high temperature properties is considered a sinter additive for hot pressing of TiB_2.[34]

Table 3: conditions for densification of boron carbide and refractory borides

Material	Sinter addition	Process parameters	Relative density %ρ_{th}	Hardness, GPa	Microstructure
B_4C D_{50}-0.5µm	Nil	PS – 2375°C	90	24-25	Coarse grains 50-120µm
	3.0 % C D_{50}-18 µm	PS – 2325°C	90	24-25	Fine grains <20 µm
	5.0 % ZrO_2 D_{50}-<1 µm	PS – 2275°C	93	32	Two phases <20 µm
	Nil	HP- 1900 °C	98	32	1-2 µm
TiB_2 D_{50}-1.1 µm	Nil	HP - 1700°C	88	6-7	Presence of pores
	Nil	HP - 1800°C	98	26	1-2 µm
	20% $MoSi_2$ D_{50}-1.4 µm	PS-1900°C	88.5	-	-
	20% $MoSi_2$	HP-1700°C	98.7	25	1-2 µm
	2.5% $TiSi_2$ D_{50}-15 µm	HP-1550°C	99	20-25	1-2 µm
	5% $CrSi_2$ D_{50}-18 µm	HP-1700°C	98.6	28	-
	2.5% CrB_2 D_{50}-4.8 µm	HP-1750°C	96.6	18	2-3 µm
ZrB_2 D_{50}-2.0µm	Nil	HP-1850°C	100	15	4 µm
	2.5% CrB_2 D_{50}-1.8 µm	HP-1750°C	100	17	2 µm
HfB_2 D_{50}-2.2µm	Nil	HP-1850°C	80	15.2	-
	5%$TiSi_2$ D_{50}-15 µm	HP-1650°C	95	23.3	-
CrB_2 D_{50}-4.8µm	Nil	PS-2000 °C	88	10	5 µm
	Nil	HP-1600°C	100	16	3-5 µm

PS- pressureless sintering, HP- hot pressing, D_{50}- mean particle diameter

The addition of 10–20 wt% $MoSi_2$ enabled to achieve 97–99%ρ_{th} in the composites at 1700°C under similar hot-pressing conditions. The densification was found to be assisted by liquid phase formation of $TiSi_2$.[35] The hot-pressed TiB_2–10 wt% $MoSi_2$ composites exhibit a Vickers hardness of 26–27 GPa and modest indentation toughness of 4–5 MPa.m$^{1/2}$. Other silicides with low melting points such as $TiSi_2$ (1540°C) and $CrSi_2$ (1477°C) have also been tested as sintering additive for consolidation of TiB_2. With a small addition of 2.5% $TiSi_2$, pellets of 99%ρ_{th} could be obtained by hot pressing at a much lower temperature of 1550°C. Hardness (20-25GPa), Fracture toughness (4.6-6.3 MPa.m$^{1/2}$) and flexural strength (390 MPa) of this composite is found to match that of monolithic TiB_2 of similar density.[36] Coefficient of Thermal Expansion (CTE) is found to be close to that of TiB_2 and electrical resistivity is slightly higher than that of TiB_2. Oxidation resistance of this composite is found to be superior to that of TiB_2+$MoSi_2$ and is close to that of TiB_2. The formation of SiO_2 layer in addition to TiO_2 on the surface is the reason for the improved oxidation resistance of this composite.[36] Hence $TiSi_2$ is found to be the best sinter additive for hot pressing of TiB_2. In the case of $CrSi_2$ as sinter additive, the optimum hot pressing temperature was found to be 1700°C. With 5% addition of $CrSi_2$ a composite of 98.6% ρ_{th} with a hardness of 28 GPa and fracture toughness of 4.85 $MPa.m^{1/2}$ was obtained. CrB_2 with a melting point of 2200°C has also been found to be a suitable sinter additive especially for improving the oxidation resistance of TiB_2. The oxidized layer was found to contain TiO_2 and $CrBO_3$. Mechanical, physical properties and oxidation data of TiB_2 and composites containing $MoSi_2$, $TiSi_2$, CrB_2 and $CrSi_2$ prepared in our laboratory by hot pressing are presented in table 4.

Table 4. Properties of hot pressed TiB_2 and composites

Material	Hardness GPa	Fracture toughness, (K_{IC}) $MPa.m^{1/2}$	Flexural strength (σ) MPa	CTE x 10^6 K^{-1}	Thermal conductivity W/m/K	Electrical resistivity $\mu\ \Omega$-cm	Slope of general rate equation for oxidation (m)*
TiB_2	26	5.1	-	-	62.5	15.15	1.414
TiB_2+20%$MoSi_2$	25	5.1	-	-	62.0	24.63	1.423
TiB_2+2.5%$TiSi_2$	18	5.3	390	5.7	-	22	1.789
TiB_2+2.5%CrB_2	18	2.8	295	6.1	-	40	1.901
TiB_2+5%$CrSi_2$	28	4.9	-	-		-	-

*$(\Delta w/A)^m = K_m.t$, Where Δw- is the change in weight, A- surface area of the sample, t- oxidation time and K_m- rate constant.

MOLYBDENUM DISILICIDE

The process chosen for preparation of $MoSi_2$ was synthesis from its elements because both the powders Mo and Si are available in pure form. The pelletized stoichiometric mixture of molybdenum and silicon was heated under 0.1 Pa to ~1400°C. The reaction is highly exothermic and instantaneous. The temperature of the charge shoots up by a few hundred degrees as the reaction starts and the reaction is completed within a few minutes. The charge pellets should be kept with sufficient space between them, and the furnace power controlled to avoid melting of the charge. As both the elements molybdenum and silicon are highly reactive, they should be stored in inert atmosphere. In presence of air/ oxide surface, loss of molybdenum is observed at above 900°C due to the formation and evaporation of MoO_3.[37] This drastically affects the recovery and also the formation of pure phase $MoSi_2$. XRD pattern of molybdenum disilicide produced in our laboratory is given in figure 2, which shows it to be pure $MoSi_2$.

ZIRCONIUM/HAFNIUM BORIDE

Synthesis of ZrB_2 also was carried out by a reaction similar to that of TiB_2. In this case, the product of reaction at all the temperatures upto 1800°C was a mixture of ZrB_2, ZrB and C. While comparing the synthesis of TiB_2 and ZrB_2 by boron carbide reduction of their oxides, the free energy change for the reaction is positive upto 970°C for TiB_2 formation and that for ZrB_2 is upto 1430°C. Hence ZrB_2 formation is possible only at higher temperature compared to TiB_2. Free energy change of the reaction for the formation of borides is given in figure 1. Hafnium and Zirconium are chemically very similar and this reflects in the formation of borides also. The formation of ZrB may be due to the loss of boron from the reaction zone by evaporation of B_2O_3 or suboxides of boron before the formation of borides. The charge composition was suitably modified with excess boron carbide to obtain a pure phase product of ZrB_2. Pure HfB_2 was obtained at a slightly higher temperature of 1825°C (table 2).

ZrB_2 powder was hot pressed to full density at 1850°C by application of 35 MPa pressure and a holding period of 2 hours in vacuum. Addition of CrB_2 lowers the sintering temperature by 100°C. CrB_2 also prevents the grain growth of ZrB_2 during hot pressing and thus improves hardness and fracture toughness. However density of HfB_2 hot pressed at 1850°C was low at 80%ρ_{th}. When 5% of $TiSi_2$ was added to HfB_2, higher density of 95%ρ_{th} was achieved at a low temperature of 1650°C. Hardness of this composite was also high at 23.3GPa (table 3).

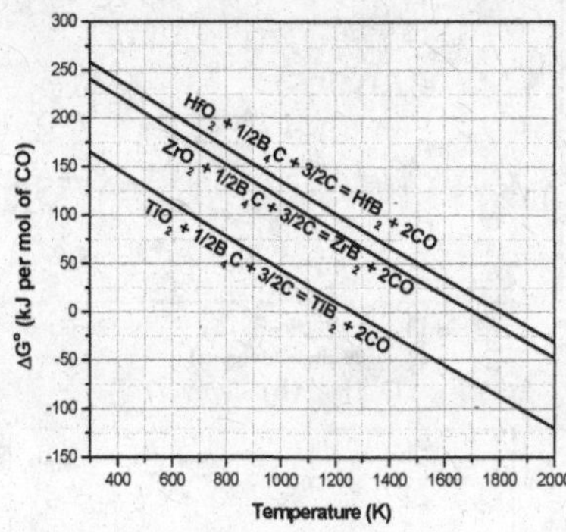

Fig.1 Free energy change in the formation of borides

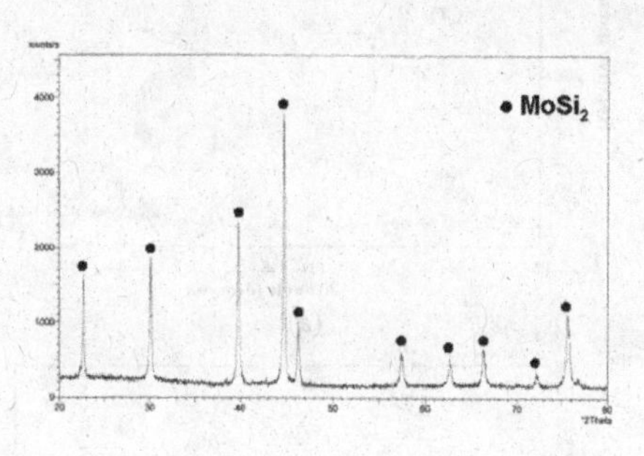

Figure 2. XRD pattern of $MoSi_2$ powder

CHROMIUM BORIDE

When synthesis of CrB_2 from a stoichiometric charge composition was carried out at 1700°C, though the weight loss was close to the theoretical value (36.3%), the product obtained was composed of CrB_2, Cr_3B_4 and graphite phases. The presence of boron deficient phase (Cr_3B_4) indicates the possible loss of boron from the charge. By using a modified charge composition containing more boron and less carbon ($Cr_2O_3 : B_4C : C = 1 : 1.2 : 1.31$), single phase CrB_2 was prepared.

By pressureless sintering CrB_2 powders at 1800°C, a maximum density of 93%ρ_{th} was obtained after holding for 6 hours. However hot pressing results in a fully dense pellet of CrB_2 at 1600°C and 35 MPa pressure in 2 hours. The hardness and fracture toughness of fully dense CrB_2 was 16 GPa and 3.5 $MPa.m^{1/2}$ respectively.

RARE EARTH BORIDES

Rare earth (RE) borides are generally synthesized by direct combination of elements, borothermic reduction of RE oxides or by carbothermic reduction of a mixture of RE oxide and boron oxide in a high temperature furnace (1500-1800°C). Borides obtained by the above methods are not pure and often difficult to obtain in precise control of stoichiometry.[38] Synthesis of Europium hexaborides (EuB_6) was attempted by boron carbide reduction method according to the following reaction.

$$Eu_2O_3 + 3B_4C \rightarrow 2EuB_6 + 3CO \quad (2)$$

Pure EuB_6 was prepared at 1400°C and a vacuum of 1 Pa. XRD pattern of the product is given in figure 3a. However, preparation of pure YB_6 by the same method has not been possible and the product obtained was always a mixture of YB_6 and YB_4 (figure 3b).

Fused salt electrolysis offers an alternative low temperature method and the possibility of obtaining pure stoichiometric compounds due to the selective nature of the process. A novel process has been developed for the preparation of RE borides by using an oxyfluoride bath[39]. Pure NdB_6 and SmB_6 have been synthesized by fused salt electrolysis at ~ 900°C. XRD patterns of these two borides are presented in figure 3c and 3d.

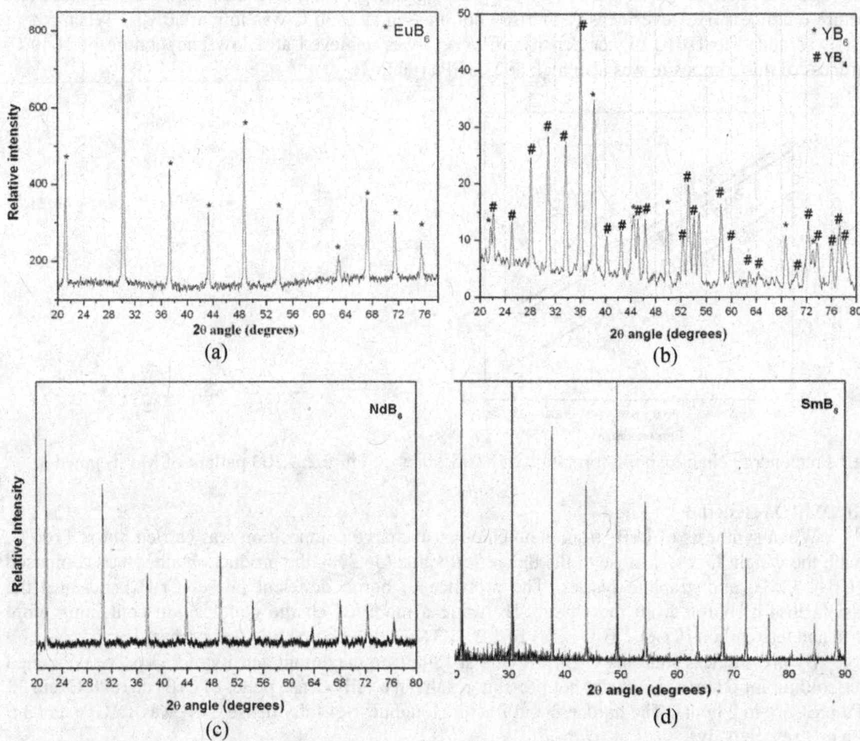

Figure 3. XRD patterns of rare earth borides a) EuB_6 b) YB_6 c) NdB_6 and d) SmB_4

CONCLUSION

Synthesis of phase pure compounds of many of the carbides, borides and silicides is challenging. A reliable method for production of B^{10} enriched boron carbide is by synthesis from the elements. Boron carbide prepared by this method in our laboratory has been consolidated by hot pressing to >95%ρ_{th}. The mechanical and physical properties of this compact have been evaluated and found suitable for use as control rod material.

In the case of refractory borides, prepared by oxide/ carbide reactions, two problems exist. During the course of the reaction, boron losses occur due to the vaporization of boron containing species formed by oxidation of B_4C or by side reactions. This is overcome by adding extra boron in the charge and controlling the furnace atmosphere to avoid oxidation. Higher temperatures and vacuum are usually needed to obtain a product low in carbon and oxygen.

In the formation of CrB_2, close control of charge composition is needed to form a pure phase. Here again, loss of boron is compensated in the charge and by carrying out the reaction in rough vacuum.

In all the cases, close contact between the particles help in achieving faster and more complete reaction. In practice, the charge constituents are milled together to obtain a fine mixture in the range of a few microns and then compacted to form pellets. As boron carbide is a very hard and abrasive material, the parts coming in contact with the powders are to be lined with a material such as WC and grinding balls also made of WC.

Due to high melting point, presence of oxide layer on the surface, low values of self diffusion and covalent nature, densification of carbides and borides are difficult. In case of pressureless sintering of B_4C, addition of carbon and ZrO_2 can be effectively used as sintering aids to obtain high densities and grain refinement. Hot pressing of borides by the addition of low melting compounds such as $MoSi_2$, $TiSi_2$, CrB_2 and $CrSi_2$ are seen to give compacts with high densities and hardness. Improved oxidation behavior was observed with the addition of $TiSi_2$ and CrB_2 by the formation of a protective oxide layer.

EuB_6 was produced in pure form by reaction between RE oxide and boron carbide at 1400°C. But YB_6 formed by the same process always contained YB_4 also. Pure SmB_6 and NdB_6 were produced by fused salt electrolysis method at a much lower temperature of 900°C.

ACKNOWLEDGEMENTS

The authors appreciate the contribution of their colleagues, Mr.T.S.R.Ch.Murthy Mr.J.K.Sonber and Mr.R.K.Fotedar who have been part of the team in carrying out the synthesis, consolidation and characterization of these materials.

REFERENCES

1. R. G. Munro, ''Material Properties of Titanium Diboride,'' J. Res. Natl. Inst. Stand. Technol., **105 [5]** 709–20 (2000).
2. B.Basu, G.B.Raju, and A.K.Suri, "Processing and properties of monolithic TiB_2 based materials", International Materials Reviews **51[6]** 354-74 (2006).
3. R. A. Cutler, ''Engineering Properties of Borides''; in Engineered Materials Handbook, Vol. 4. Ceramics and Glasses. Edited by S. J. Schneider Jr. ASM International, Metals Park, OH, 787–803 (1991).
4. Matkovich V. I. Boron and refractory borides. New York: Springler-Verlag; 1977
5. Chamberlain AL, Fahrenholtz WG and Hilmas GE. High strength zirconium diboride based ceramics. J Am Ceram Soc., **87(6)**, 1170-1172 (2004).
6. J. Castaing, P. Costa, Properties and Uses of Diborides, Springer-Verlag, 1977.
7. C. C. Klepper: "Sintered boron as high-strength lightweight structural material for aerospace vehicles", (http://optics.nasa.gov/tech_days/tech_days_2005)
8. A.Lipp, "Boron Carbide: production properties and application", Technische Rundschau, **14, 28, 33** (1965) and **7** (1966).

9. F.Thevenot: "Boron Carbide-A comprehensive review", Journal of European Ceramic Society, **6**, 205–225 (1990).
10. R. Jimbou, M. Saidoh, K. Nakamura, M. Akiba, S. Suzuki, Y. Gotoh, Y. Suzuki, A. Chiba, T. Yamaki, M. Nakagawa, K. Morita and B. Tsuchiya, "New composite composed of boron carbide and carbon fiber with high thermal conductivity for first wall", J. Nucl. Mater., **233-237**, 781-786 (1996).
11. P. Valentine, P. W. Trester, J. Winter, J. Linke, R. Duwe, E. Wallura and V. Philips, "Boron carbide based coatings on graphite for plasma-facing components", J. Nucl. Mater., **212-215**, 1146-1152 (1994).
12. R. Jimbou , K. Kodama, M. Saidoh, Y. Suzuki, M. Nakagawa, K. Morita and B. Tsuchiya, "Thermal conductivity and retention characteristics of composites made of boron carbide and carbon fibers with extremely high thermal conductivity for first wall armour", J Nucl. Mater., **241-243**, 1175-1179 (1997).
13. J. G. V. D. Laan, G. Schenedecker, E. V. V. Osch, R. Duwe and J. Linke, "Plasma sprayed boron carbide coating for first wall protection", J. Nucl. Mater., 1994, **211**, 135-140.
14. Fahrenholtz WG and. Hilmas GE. Refractory diborides of zirconium and hafnium, J. Am. Ceram. Soc; **90(5)**1347-64 (2007)
15. Opeka M. M,. Talmy I. G and Zaykoski J.A, "Oxidation-based materials selection for 2000°C + hypersonic aerosurfaces: Theoretical considerations and historical experience", J. Mater. Sci., **39,** 5887-904 (2004).
16. Levine S.R, Opila E.J, Halbig M.C, Kiser J.D, Singh M and Salem J.A, "Evaluation of ultra-high temperature ceramics for aeropropulsion use", J. Euro. Ceram. Soc., **22**, 2757-67 (2002).
17. L.R.Jordan, A.J.Betts, K.L.Dahm, P.A.Dearnley, G.A.Wright, "Corrosion and passivation mechanism of chromium diboride coatings on stainless steel", Corrosion Science **47**, 1085-1096 (2005).
18. X.H.Zhang, P Hu, J.C.Han, L.Xu and S.H.Meng, "The addition of lanthanum hexaboride to zirconium diboride for improved oxidation resistance", Scipta Materialia **57**, 1036-1039, (2007).
19. C.E.Holcombe, Jr., L.Kovach, "Stable materials for fluorine-rich environments", Ind.Eng.Chem.Prod.Res.Dev., **21**, 673-676 (1982).
20. R. Mitra, "Mechanical behaviour and oxidation resistance of structural silicides", International Materials Reviews, **51[1]**, 13-64 (2006)
21. M. Bouchacourt and F. Thevenot, "The properties and structure of the boron carbide phase", J. Less Comm. Met., **82**, 227-235 (1981).
22. D. Gosset and M. Colin, "Boron Carbides of various compositions: An improved method for x-ray characterisation", J. Nuclear Mater., **183**, 161-173 (1991).
23. M. Bouchacourt and F. Thevenot, Journal of the less common metals, **82**, 219-226, (1981).
24. G. Goller, C. Toy, A. Tekin and C. K. Gupta, "The production of boron carbide by carbothermic reduction", High Temperature Materials and Processes, **15[1-2]**, 117-122 (1996).
25. A. Aghai, C. Falamaki, B. E. Yekta and M. S. Afarani, "Effect of seeding on the synthesis of B_4C by the magnesiothermic reduction route", Industrial ceramics, **22 [2]**, 121-125 (2002).
26. E.G.Gray, "Process for the production of boron carbide", US patent No. 2,834,651 (1958).
27. B. Chang, B.L. Gersten, S.T. Szewczyk and J. W. Adams, "Characterization of boron carbide nanoparticles prepared by a solid state thermal reaction", Applied physics A, **86,** 83-87 (2007).
28. S. T. Benton and R. David, "Methods for preparing boron carbide articles", US patent No. 3,914,371 (1975).
29. T.K.Roy, C. Subramanian and A.K.Suri, "Pressureless sintering of boron carbide", Ceramics International, **32**, 227-233 (2006).

30. C.Subramanian, T.K.Roy, T.S.R.Ch.Murthy, P.Sengupta, G.B.Kale, M.V.Krishnaiah and A.K.Suri, "Effect of Zirconia addition on pressureless sintering of boron carbide", Ceramics International, **34**, 1543-1549 (2008).
31. H. Lee and R. F. Speyer, "Pressureless sintering of boron carbide", J.Am.Ceram.Soc., **86[9]**, 1468-73 (2003).
32. R. F. Speyer and J. Lee, "Advances in pressureless densification of boron carbide", Journal of material science, **39**, 6017-6021 (2004)
33. C. Subramanian, T. S. R. Ch. Murthy and A. K. Suri "Synthesis and consolidation of titanium diboride", Int. J. Refract. Met. And Hard Mater., **25**, 345-350 (2007).
34. T. S. R. Ch. Murthy, B. Basu, R. Balasubramanian, A. K. Suri, C. Subramanian and R. K. Fotedar "Processing and properties of TiB_2 with $MoSi_2$ sinter additive: A first report", J. Am. Ceram. Soc., **89[1]**, 131-138 (2006).
35. K. Biswas, B.Basu, A.K.Suri, K.Chattopadhyay, "A TEM study on TiB_2-20%$MoSi_2$ composite: Microstructure development and densification mechanism", Scripta Materialia, **54**, 1363-1368 (2006).
36. T.S.R.Ch.Murthy, C.Subramanian, R.K.Fotedar, M.R.Gonal, P.Sengupta, Sunil Kumar, A.K.Suri, **"**Preparation and property evaluation of TiB_2+$TiSi_2$ composite**"** International Journal of Refractory Metals and Hard Materials; doi10.1016/j.ijrmhm.2008.10.001 (2008)
37. Yuntian T. Zhu, Marius Stan, Samuel D. Conzone and Darryl P. Butt, "Thermal Oxidation Kinetics of MoSi2-Based Powders**"** *J. Am. Ceram. Soc.,* **82 [10],** 2785–90 (1999)
38. A Latini, F D Pascasio, D Gozzi, "A new synthesis route to light lanthanide borides: borothermic reduction of oxides enhanced by electron beam bombardment", Journal of alloys and compounds, **346**, 311-313 (2002).
39. L.John Berchmans, A.Visuvasam, S. Angappan, C.Subramanian and A.K. Suri, **"A** novel electrochemical synthesis route for Rare Earth hexaborides**",** (Unpublished work).

SHRINKAGE REDUCTION OF CLAY THROUGH ADDITION OF ALUMINA

K. Hbaieb
Institute of materials research and engineering (IMRE)
3 Research link, Singapore 117602

ABSTRACT

In this paper the packing density, shrinkage, rheological and mechanical properties of clay/alumina mixtures are investigated. The packing density and shrinkage of clay during drying and firing is improved by adding alumina. Consolidated bodies formulated from mixtures of different amounts of clay and alumina exhibited plastic behavior during uniaxial compression. Flow behavior of particles in the slurry state was governed by the strongest third network- formed through the contact of alumina and clay particles- when both alumina and clay powder have similar particle sizes. For mixtures where large and small particles are in interaction the viscosity behavior of the slurry depend on the volume fraction of both constituents.

INTRODUCTION

Since clay usually packs at too low density, cracking and high residual stresses might arise during drying and densification. In the old days, the Chinese have found out that if they mix clay with a filler along with a plasticiser they can usually maintain the plasticity and malleability of clay as well as reduce shrinkage, thus overcome cracking problems during densification. The plasticiser used was usually feldspar, whereas quartz was the filler. Since quartz is very hard and rigid it undergoes much less shrinkage during densification and thus constrains tremendously shrinkage of clay particles.

Lange et al.[1] have been working for the last decade on improving the mechanical behavior of alumina ceramics using colloidal processing. By controlling interparticle pair potential they proved that they can manipulate the mechanical and rheological behavior of alumina powder compacts; interestingly, they showed that alumina powder compacts can deform plastically and thus can be shaped into complex shapes by extrusion methods. Uniaxial compression tests showed that consolidated bodies formulated from pure alumina slurries exhibited clay like behavior (low flow stress and high strain of failure). The packing densities of these bodies were around 0.6, which is much higher than clay (around 0.4). In addition, during densification alumina bodies exhibit much less shrinkage than clay. Hence, we expect that mixing clay with alumina would maintain the plastic characteristic of clay, increase the packing density and also reduce shrinkage upon heating to high temperatures.

In this paper we present a study of mixing different alumina powder containing different particle sizes with clay. Rheological and mechanical behaviors of these mixtures are investigated. In addition, shrinkage during drying and densification is characterized.

EXPERIMENTAL METHOD

Slurry preparation

Slurries were prepared by mixing alumina powder (Sumitomo Chemical Company, New York, AKP-50 Grade, d50 ≈ 0.25 microns, AKP-15 Grade, d15 ≈ 0.6 microns, AA-2 Grade, d2 ≈ 2 microns and AA-5 Grade, d5 ≈ 5 microns) or/and clay powder (Huber, Polygloss 90, d90 ≈ 0.5 microns) with dionized water at a solid volume fraction of 0.2. The slurry was then dispersed by untrasonication. In slurries using AKP-50 as the principal solid KOH was added to increase pH to high values (around 11). This produces negative charges on the particles, resulting in a repulsive interparticle pair potential and thus a dispersed slurry. Tailoring surface charge was not necessary when using the other types of powder since their large particle size implies a lower number per unit volume and a smaller surface area, thus no strong attractive force is active. It is desirable to form a weakly attractive potential that allow particles to be at very close proximity to each others yet not touching, thus a very strong non touching network can be formed with low flow stress yet high packing density. To produce the desired interparticle potential, i.e. a weakly attractive network where the particles are separated by a small distance yet not touching, the slurry was coagulated by adding 0.1 M KCl salt. The pH was then changed to 7.5 to optimize the effect of the interparticle potential. Finally the slurries were stored in a roller for several hours to stabilize.

Consolidation and packing density determination

Consolidation of slurries were made through the pressure filtration method, which is described elsewhere[2]. The slurry was poured in one-inch diameter die cavity. Constant pressure was applied with a plunger until an equilibrium density was reached. Once the plunger ceased moving for a certain time, usually not less than 10-15 min., pressure filtration was considered complete. After the pressure was released, the consolidated body was removed, weighted and stored in a plastic bag containing a wet paper to avoid drying. Density was determined as follows: The consolidated body was heated to 70 ℃ for -at least- 20 hours to remove water. The body was then again weighed to determine the weight loss, which was assumed to be completely due to water removal. To eliminate residual salt, the consolidated body was then heated to 350 ℃ and then weighed. Thus, the density can be calculated knowing the weight loss due merely to water and salt, respectively.

Mechanical testing and shrinkage measurement

Mechanical testing was performed using a screw-driven mechanical test frame (Instron model 8562, Canton, MA). Saturated, cylindrical-shaped specimens contained in a plastic bag were subjected to uniaxial compression. Strain rate was kept at a constant magnitude of 1 mm/min. Load-displacement data was taken by a computer connected to the machine. Engineering vs. strain curves were then computed from the load-displacement data taking into account the following: Strain was calculated by dividing the measured displacement by the initial thickness of the specimens. The current area was determined assuming that the volume of the specimen was preserved and the effect of barreling due to friction between specimen and loading platens was insignificant. Stress was then calculated by dividing the measured load by the current area.

After the load-displacement data was taken the dimensions of the specimens were measured before and after heating the specimens to 70 ℃. Shrinkage was then calculated from these measurement data.

TGA analysis and weight loss measurement

TGA analysis was performed on Polygloss powder using Netzsch apparatus (Mode STA 409). It was found that the powder lost around 12% of its weight in a temperature range between 450 and 600 ℃. Above 650 ℃ no more weight loss was detected. Consolidated bodies formulated from slurries containing mixtures of AKP-15/ Polygloss and AA-5/ Polygloss powder were fired at 700 ℃. Shrinkage as well as weight loss was then calculated. The experimentally measured weight loss was compared to theoretical calculation. Note that water loss calculated by this experiment is the water intercalated between the layers in the structure itself. Water that was flowing around the particles in the consolidated body was completely eliminated when heating the body at 70 ℃.

Viscosity measurement

Slurry viscosity was measured using a dynamic stress rheometer (Rheometrics DSR) with a couette-type measurement cell (29.5 mm diameter, 44.0 mm long). Slurries at 0.2 volume fraction were subjected to high stresses which were decreased until the shear rate reached values too low for accurate measurement.

RESULTS

Mechanical Behavior

Before looking at the effect of introducing alumina powder, consolidated bodies made from pure clay (Polygloss) were first investigated. The slurries were prepared at two pH values, 9 and 7.5, and by adding different amounts of salt precisely 0.02, 0.05 and 0.1 M KCl. Consolidation of these slurries was performed at a constant pressure of 3 MPa. The resulting compact bodies were then tested in uniaxial compression at a constant strain rate of 1mm/min. Test results are summarized in Table 1, and show that increasing the amount of salt results in decreased packing density and flow stress. Similar results on slurries containing alumina powder were reported by Franks *et al.*[3] They showed that adding salt causes a reduction in interparticle pair potential as well as a decrease in packing density. In addition the slope of the interparticle potential versus interparticle distance becomes steeper the more salt is added.

Table 1: Change in flow stress and packing density with adding salt

pH	Salt Content	Packing Density	Flow Stress in MPa
7.5	0.02 M	55.0 v.%	1.7
7.5	0.05 M	52.0 v.%	1.1
7.5	0.10 M	45.4 v.%	0.4
9	0.02 M	57.7 v.%	2.2
9	0.05 M	52.0 v.%	1
9	0.10 M	47.0 v.%	0.5

Holding the volume fraction of solids at a constant 20%, slurries with three different mixtures of alumina powder and Polygloss were prepared using the same procedure previously discussed. The volume of alumina was the same, one third or two third of the volume of Polygloss powder. 0.1 M KCl was added to each slurry. PH and consolidation pressure were held at 7.5 and 3 MPa, respectively. The packing density was highest when the volume of small particles (AKP-50) was one third the amount of large particle (clay), i.e. when the volume fraction of small particles relative to the total solid content was 0.25. When the alumina particles are larger than the clay (AA-5, AA-2 and AKP-15) the packing density increases with increasing alumina content. For the consolidated bodies formulated from slurries containing mixture of AKP-15/Polygloss powder the packing densities were effectively independent of the relative amounts of each powder. Knowing that AKP-15 and Polygloss particles have similar particle sizes, different mixtures of the two should have similar packing densities. The data showing the packing densities of the consolidated bodies are summarized in Table 2.

Table 2: Packing density and flow stresses of consolidated bodies

Packing density	Clay vol. %	Alumina vol. %	Flow stress in MPa
Mixture of Clay/AA-5			
57.6 v.%	0%	100%	0.05
69.8 v.%	25%	75%	0.3
63.4 v.%	50%	50%	0.5
54.8 v.%	75%	25%	0.5
45.4 v.%	100%	0%	0.4
Mixture of Clay/AA-2			
	0%	100%	0.05
61.0 v.%	25%	75%	0.35
57.0 v.%	50%	50%	0.43
51.0 v.%	75%	25%	0.45
45.4 v.%	100%	0%	0.4
Mixture of Clay/AKP-15			
	0%	100%	0.05
52.6 v.%	25%	75%	0.2
50.8 v.%	50%	50%	0.3
49.9 v.%	75%	25%	0.45
45.4 v.%	100%	0%	0.4
Mixture of Clay/AKP-50			
51.0 v.%	0%	100%	0.2
43.3 v.%	25%	75%	0.2
46.7 v.%	50%	50%	0.225
47.0 v.%	75%	25%	0.4
45.4 v.%	100%	0%	0.4

A typical stress-strain curve for consolidated bodies tested in uniaxial compression is shown in figure 1. Table 2 shows the flow stresses computed from the different experiments performed on the consolidated bodies. These results show that the flow stress of alumina was always lower than the flow stress of Polygloss. The flow stresses of consolidated bodies formulated from slurries of alumina/Polygloss mixtures were always higher than the pure alumina. This was thought to be result of the three following reasons: Alumina and Polygloss particles have different surface charges (at pH=7.5); Zeta potential measurements show that alumina particles have positive surface charges while Polygloss particles have in contrast negative charges at their surfaces. Adding salt causes counterions to coat the particles in an attempt to neutralize the surface. Since the amount of salt added is low (only 0.1 M KCl) effective surface charges on both alumina and Polygloss particles remain still intact. These opposite charges introduce a third force not familiar in the pure slurry of one type of particles. Since this force is negative the interaction of alumina and Polygloss particles produces an attractive network that prevents the ease flow of alumina particles past Polygloss particles. Moreover, some of the mixtures have a packing density higher than that of pure alumina and this has an impact of increasing flow stress. A third reason for the increase of flow stress of the mixture system relative to pure alumina is thought to be the high flow stress of Polygloss consolidated body. Introduction of Polygloss powder to the alumina slurry strengthens the alumina network and contribute to the increase of flow stress.

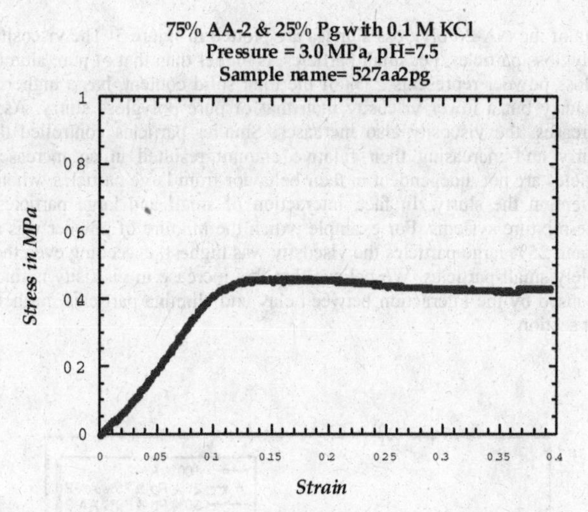

Figure 1: *typical stress/strain curve for consolidated bodies tested in compression.*

Rheology Behavior

Figure 2 shows the rheology results for the mixture of AA-5/polygloss. The viscosity of small polygloss particles is clearly higher than the viscosity of large AA-5 particles. The viscosity of the mixture is in-between these two values. As we increase the volume fraction of small polygloss particles, the viscosity also increases.

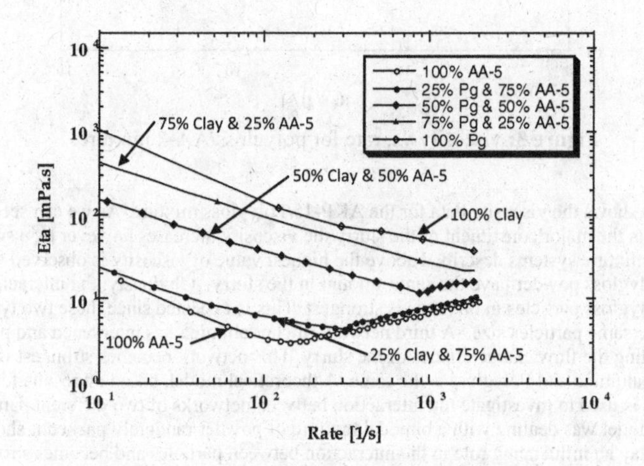

Figure 2*: Viscosity vs. Rate for polygloss/AA-5 mixture.*

Viscosity data of the AA-2/Polygloss slurries are plotted in figure 3. The viscosity of the slurry containing 100% polygloss particles, i.e. small particles, is higher than that of pure alumina slurry. The slurry, where polygloss powder represents 25% of the total solid content, has a higher viscosity than that of pure AA-2 slurry but a lower viscosity than that of pure polygloss slurry. As the polygloss powder content increases, the viscosity also increases. Smaller particles controlled the rheological behavior of the slurry and increasing their relative amount resulted in an increase in viscosity. However, small particles are not independent in their behavior from large particles, when the latter had relatively large content in the slurry. In fact, interaction of small and large particles is obviously observed in the three mixture systems. For example, when the mixture of powder was made up from 75% small particles and 25% large particles the viscosity was highest, exceeding even the value for the slurry containing solely small particles. We believe that this increase in viscosity is due to the strong attractive network caused by the interaction between clay and alumina particles; further discussion is presented in the next section.

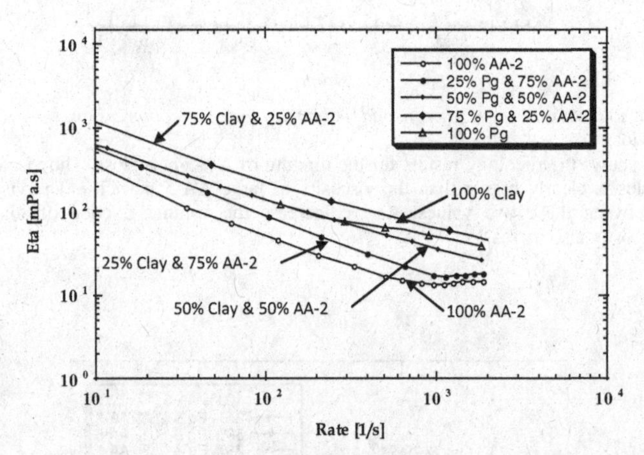

Figure 3: viscosity vs. rate for polygloss/AA-2 mixture.

Figure 4 shows the viscosity data for the AKP-15/Polygloss mixture. As we can see, when one type of powder is the major constituent of the slurry the viscosity increases however by a small amount. Unlike, for the mixture systems described above the highest value of viscosity is observed when alumina and Polygloss powder have the same amount in the slurry. Obviously, the interaction of alumina and Polygloss particles in this case is strongest. This is expected since these two types of powder have the same particles size. A third network of clay-alumina has intervened and played a big role in determining the flow characteristics of the slurry. This network becomes strongest when the amount of both alumina and Polygloss is the same. A theoretical model, based on results found by Bouvard *et al.*[3], is used to investigate the interaction between networks of two different particles. Although this model was dealing with a bimodal mixture of powder randomly packed it shows that a third network play an influencing role in the interaction between particles and becomes strongest when the two different particles have the same volume fraction.

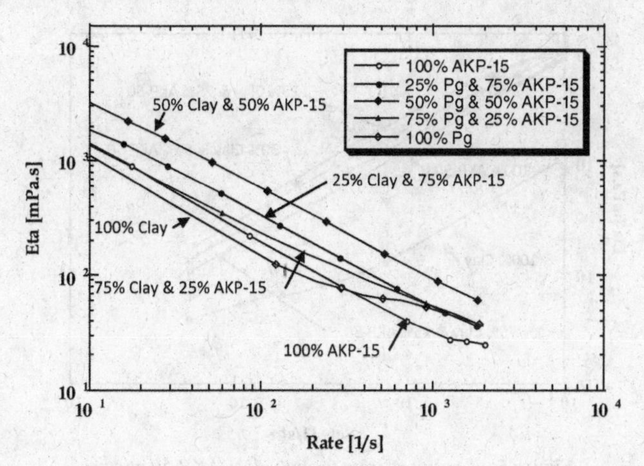

Figure 4 *viscosity vs. rate for polygloss/AKP-15 mixture.*

The last mixture system tested for rheology behavior analysis was the AKP-50/polygloss mixture (figure 5). Here the mixture exhibits higher viscosity than either pure alumina or polygloss slurry independent from the relative amounts of both alumina and polygloss powder. A heterogeneous strong attractive network formed by interaction between polygloss and alumina particles must impart this increase in viscosity. Note, that the viscosity increases with increasing the volume fraction of small AKP-50 particles.

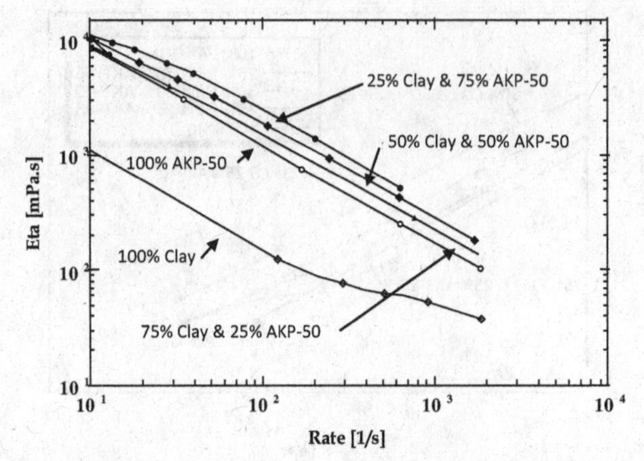

Figure 5: *viscosity vs. rate for polygloss/AKP-50 mixture.*

Shrinkage upon drying

Shrinkage measurements upon drying were performed for AKP-15/Polygloss, AA-2/Polygloss and AA-5/Polygloss mixtures. Shrinkage data are illustrated in table 3. Shrinkage is reduced when the packing density increases. Unlike for the system AKP-15/Polygloss where alumina and clay particles have similar sizes, shrinkage decreases tremendously for the AA-2/Polygloss and AA-5/Polygloss systems, when the volume fraction of alumina relative to the total solid content increases. Since alumina particles are relatively larger than Polygloss particles introduction of alumina particles increases the packing density and thus reduces shrinkage. Investigation of the AKP-50/Polygloss mixture is not of an importance in this case since introduction of alumina particles to Polygloss slurry would not lead to significant increase of packing density and therefore reduction of shrinkage for this system can not be thoroughly optimized.

Table 3: Shrinkage measurement upon drying

Packing density	Clay vol. %	Alumina vol. %	Radial shrinkage
Mixture of Clay/AA-5			
69.8 v.%	25%	75%	-1.2%
63.4 v.%	50%	50%	-2.4%
54.8 v.%	75%	25%	-3.0%
Mixture of Clay/AA-2			
61.0 v.%	25%	75%	-0.6%
57.0 v.%	50%	50%	-1.4%
51.0 v.%	75%	25%	-2.2%
Mixture of Clay/AkP-15			
52.6 v.%	25%	75%	-1.3%
50.8 v.%	50%	50%	-1.4%
49.9 v.%	75%	25%	-2.2%

Weight loss and shrinkage upon firing

Weight loss of water due to calcining of alumina/Polygloss mixture was investigated. Only two mixtures systems, AKP-15/Polygloss and AA-5/Polygloss, were examined. Table 4 shows the results detected for these two systems. It can be seen, that increasing the volume fraction of Polygloss results in an increase in weight loss. Theoretical calculations showed very good agreement with the experimental measurements and established the fact that weight loss was completely due to water removal from the particle structure itself. Accompanied with weight loss determination, shrinkage of the specimen was also investigated. In this case shrinkage was not related to change in packing density but rather to the volume fraction of Polygloss in the mixture system (Note that this was also true for the AKP-15/Polygloss mixture where alumina and Polygloss particles have similar sizes). When the volume of clay increases shrinkage also increases which also shows that water removed from the body is due to the structural water in the particles themselves and not to water flowing around particles.

Table 4: Weight loss and shrinkage upon firing

Clay vol. %	**Alumina vol. %**	**Measured weight loss**	**Theoretical weight loss**	**Radial shrinkage**	**Axial shrinkage**
Mixture of Clay/AA-5					
25%	75%	0.60 g	0.58 g	-0.52 %	-0.38%
50%	50%	1.16 g	1.09 g	-0.90 %	-0.70%
75%	25%	1.74 g	1.63 g	-1.64 %	-1.12%
Mixture of Clay/AKP-15					
25%	75%	0.45 g	0.43 g	-0.30 %	-0.00%
50%	50%	0.85 g	0.84 g	-0.82 %	-0.65%
75%	25%	1.21 g	1.20 g	-1.00 %	-1.85%

DISCUSSION

Decreasing flow stress with increasing salt content was also observed by Frank *et al.*[4] when investigating pure alumina slurries. They showed that the long range repulsive interparticle pair potential existing when the slurry is dispersed could be reduced by adding salt. When salt is added counter-ions are attracted to the particles that have opposite surface charges. When counter-ions surround the particles the effective charge of the latter is reduced; the large repulsive force is therefore reduced. The more salt is added the more surface charges are neutralized and the less effective is the "double layer repulsive force". Besides the reduction of the repulsive pair potential range the slope of the potential curve versus interparticle distance steepens the more salt is added. Steeping of the curve of the potential versus interparticle distance means that the slope (which is a force) increases; this force counteracts the force applied by the external force during pressure filtration to push particles over the repulsive barrier deep in the primary potential well. The more salt is added the lower the number of particles that are pushed during pressure filtration. The fraction of particles pushed together would increase the packing density of the consolidated body. During uniaxial compression a peak stress is required to break the strong touching network formed during consolidation. Therefore, the more salt is added the lower is the peak stress.

Zok F. *et al.*[5] has investigated the packing density of a particulate system when particles of different sizes are mixed together. He showed that the highest packing density would result from mixing 75% of large particles with 25% of small particles. Since randomly packing of large particles would leave 36 vol.% of empty space between particles, introduction of small particles will fill 64% of this remaining free space to give the highest density of 87 vol.% when the volume fraction of large particles is 0.75. This is true when the size ratio of large particles to small particles is high. When small particles are comparable in size to large particles the region right at the contact between two touching particles has to be excluded from being filled with small particles. This concept can be well applied to our different mixture systems. For the AA-5/Polygloss mixture where the particle size ratio is highest the packing density is effectively highest when compared with other mixture systems. As the size ratio is reduced the packing density for a corresponding type of mixture is also reduced implying that more volume is excluded (i.e. the region at the contact of two touching particles) despite the presence of small particles that fill the otherwise available space.

Viscosity of bimodal mixture was investigated by holding the total volume fraction constant. Previous studies on bimodal and multimodal systems were performed by adding different amounts of large particles to a slurry of fine particles. Fidleris and Whitmore[6] showed that when the size ratio of fine particles to large particles is less than 1/10 the large particles added to a slurry of fine particles feel the same resistance as when they are in a suspension of pure liquid with the same viscosity as the slurry of fine particles. Based on these results Farris[7] predicted the viscosity of multimodal mixture from the viscosity of unimodal suspension assuming that small particles have zero interaction with large particles as long as the size ratio is less than 1/10. He also showed that adding fine particles to large particles decreases at some stages the viscosity of the suspension. Since for our case the size ratio of small to large particles is large than 1/10 we can effectively deduce that small particles definitively interact with large particles. This interaction depends on the size ratio.

When the size ratio is larger than 1/10 but less than 1 the fine particles have the largest contribution to the viscosity. When the size ratio is larger than 0.3 which is the case of the Polygloss/AA-2, Polygloss/AKP-15 and Polygloss/AKP-50 mixtures the viscosity of the mixture even exceeded the value of pure slurry of fine particle. Since the viscosity of large particles is smaller than that of the fine particles an interaction of large particles with small particles has to be the reason for this increase. A theoretical model was developed based on previous studies performed by Dodds[8] and Bouvard[9] to study the strength and volume fraction of three different networks existing in a randomly packed particulate system. In our case, these three networks are formed through the contacts of clay/clay, alumina/alumina and clay/alumina particles. This model suggests that when the volume fraction of alumina is equal to the volume fraction of clay the coordination number of alumina to clay particles is equal to three and the volume fraction of the third network, i.e. clay/alumina network, is the largest and equal to 0.5. The formed third network is strong attractive, since as already mentioned above, the effective charges of alumina and polygloss particles (at pH=7.5) are slightly affected by adding salt, consequently still remain positive and negative, respectively. Therefore, when alumina and Polygloss particles are comparable in sizes (Polygloss/AKP-15) the interaction between the different particle types is strongest among all the mixture systems and has the largest effect when the volume fractions of both types of powder are equal. Based on this theoretical thought we believe that a third network formed through the contacts of alumina and clay particles has the most contribution to viscosity of the slurry (when both alumina and clay particles are comparable in size). Thus the viscosity of Polygloss/AKP-15 mixture reaches the largest value when the mixture is made up of equal volume fractions of both alumina and Polygloss.

Since the size of polygloss particles is only double the size of AKP-50, the interaction between alumina and clay particles is, also in this case, strong. A strong attractive network is, therefore, formed due to interaction between alumina and clay particles. The strength of this network is manifested by the increase of viscosity exceeding the value of AKP-50. Unlike for the mixture of AKP-15/polygloss where the highest value for viscosity was when both alumina and polygloss have same volume fraction (50%), the highest value for the mixture AKP-50/polygloss is when the large particles constitute 25% of the total solid content. Note that AKP-15 and polygloss particles have similar sizes and therefore also similar viscosity. The size of AKP-50, however, is half the size of polygloss. Because the viscosity of polygloss is lower than that of pure AKP-50 the mixture of 75% AKP-50 with 25% polygloss exhibits higher viscosity than it would for the case when the third network is strongest (i.e. when alumina and polygloss have same volume fractions). The exactly opposite scenario is observed in the AA-2/Polygloss mixture; the mixture of 75% Polygloss with 25% AA-2 exhibits the highest viscosity. The interaction in the Polygloss/AA-5 mixture is so weak that it hardly affects the viscosity of the slurries. Since the viscosity of AA-5 is much lower than that of Polygloss, adding Polygloss to AA-5 alumina increase the viscosity of the mixture. The more clay is added the higher is the increase in viscosity.

CONCLUSION

Shrinkage of clay could be effectively reduced and therefore cracking during drying and densification could be avoided. The flow behavior of the mixture is not much different from pure clay. Along with preserving the malleability and plasticity of clay the consolidated bodies formulated from the mixtures show very high packing density especially when small and large particles make up the total solid content. The particle size ratio has influenced in some way the rheology of some mixture. This is due to interaction between alumina and clay and the formation of a third (alumina/clay) attractive network. In the case, where alumina and clay particles have similar particle sizes (Polygloss/AKP-15) a third network of alumina/clay is the strongest and has the largest contribution to the viscosity behavior. When small amount (e.g., 25%) of one type of sufficiently large particles (Polygloss/AKP-50 and AA-2/Poloygloss) were added to the slurry of small particles the viscosity was the highest. When the particles added become too large (AA-5/Polygloss) the viscosity is still high but lower than that of pure fine particles.

REFERENCES

[1]F. F. Lange, *Curr. Opin. Solid State Mater. Sci.*, **3,** 496 (1998).
[2]G. V. Franks and F. F. Lange, *J. Am. Ceram. Soc.* **79,** 3161 (1996).
[3]D. Bouvard and F. F. Lange, *Acta metall. mater.*, **39**, 3083 (1991).
[4]G. V. Franks, Ph.D. thesis (1997)
[5]F. Zok and F. F. Lange, *J. Am. Ceram. Soc.* **74,** 1880 (1991).
[6]V. Fidleris and R.N. Whitmore, *Rheological acta,* **1**, 573 (1961).
[7]R. J. Farris, *Tran. soc. of rheol.*, **12,** 281 (1968).
[8]F. F. Lange, L. Atteraas and F. Zok, *Acta metall. mater.*, **39,** 209 (1991).
[9]J. A. Dodds , *J. Coll. Inter. Sci.*, **77,** 317 (1980).

[illegible]

[illegible]

[illegible]

ITO THIN FILM COATINGS OBTAINED FROM DEVELOPED CERAMIC ROTARY SPUTTERING TARGETS

E. Medvedovski*, C.J. Szepesi
Umicore Indium Products
50 Sims Ave., Providence, RI 02909, USA

P. Lippens
Umicore Thin Film Products
p/a Kasteelstraat 7, B-2250 Olen, Belgium

K. Leitner, R. Linsbod
Umicore Materials AG
Altelandstrasse 8, FL-9496, Balzers, Liechtenstein

A. Hellmich, W. Krock
Applied Materials GmbH, Display Products Group
D-63755 Alzenau, Germany

ABSTRACT

Indium tin oxide (ITO) transparent conductive thin films are widely used as electrode layers in optoelectronic devices, such as flat panel displays, solar cells, touch panels, and some others, and the requirements for these films in terms of their quality and manufacturing efficiency are constantly growing. The use of ceramic sputtering targets of a rotary design allows to ignite plasma without arcing and to provide stable sputtering at increased power levels, which significantly enhances sputtering efficiency and reduces film processing cost. High quality films are obtained due to reduced re-deposition onto the target surface during sputtering compared with planar targets. Rotary targets with a total length up to 3 m consisting of almost fully dense (up to 99.5% of TD) hollow cylindrical ITO ceramic bodies and their technology have been developed and successfully implemented in production. The DC magnetron sputtering process using ITO ceramic rotary sputtering targets has been optimized. Thin film morphology and properties have been studied as a function of sputtering conditions. Due to high quality ceramics in rotary targets and optimized sputtering conditions, nanosized nanocrystalline ITO thin films with specific electrical resistivity as low as 180 μOhm.cm and transmittance as high as 93% (at 550 nm after subtraction of glass substrate) have been attained.

INTRODUCTION

Highly transparent and electrically conductive oxide (TCO) thin films are widely used as electrode layers in optoelectronic devices, such as flat panel displays (FPD), e.g. liquid crystal displays (LCD), organic light-emitting diodes (OLED), plasma-display panels (PDP), touch panels, electrochromic devices, antistatic conductive films and in low-emissivity coatings [1-5]. Such TCO thin films have had a great interest in photovoltaic applications for the formation of flexible thin film solar cells (e.g. based on silicon, $CuInGaSe_2$, CdTe). The films are commonly produced by conventional DC magnetron sputtering onto glass or polymer substrates, requiring a fine-tuned deposition process and high quality ceramic sputtering targets. Indium tin oxide (ITO) is the most reliable material, among various semiconducting oxide ceramics, which may be used for TCO films processing, because it provides highly homogeneous nanostructured transparent (greater than 90% of transmittance in optical range) and electrically conductive thin films in a thickness range of 50-250 nm [1-5]. The demand of ITO products (ceramics and sputter deposited films) is continuously increasing with more stringent requirements for film quality and process efficiency. Currently used planar sputtering targets in thin film processing, which consist of one or several ceramic tiles bonded onto a metallic backing plate, have some serious disadvantages. Their fast erosion in a local racetrack pattern during sputtering limits target utilization to values in the range of 25-40%. Moreover, the risk of formation of nodules on the surface of the targets, which finally leads to a disruption of the film deposition, is relatively high; thus, the need for system maintenance during target lifetime, which reduces system up-time, is inevitable using a planar cathode configuration.

In order to increase both utilization of expensive ITO ceramic sputtering targets and the efficiency of the sputtering process, rotary sputtering cathodes equipped with tubular ceramic targets should be used. Opposed to planar targets, almost the entire surface of rotary targets becomes a working surface that results in a significant increase of target utilization (typically 70-90% depending on the magnet array, i.e. several times greater than the utilization of planar targets). Erosion of the quasi entire surface area also avoids re-deposition on the target surface, so the growth of nodules can be avoided or, at least, strongly suppressed. Moreover, because the thermal load is on the entire cylindrical target surface area with rotary magnetrons, higher power loads can be applied on rotary targets. This normally leads to higher deposition rates. All these features lead to a significant reduction of the thin film production cost.

Although ceramic sputtering targets provide higher performance compared to metallic targets, ceramic rotary sputtering targets have not been widely implemented and used in the FPD and photovoltaic industries until recent time. It was especially related to ITO ceramic rotary targets, which were not existed due to difficulties to produce high density ITO large size hollow cylindrical bodies with sufficient wall thickness without shape deformation and cracking. Umicore Indium Products (UIP) was one of only a few companies that took up a challenge to develop and commercialize rotary sputtering targets consisting of high density (99% of TD or greater) ITO ceramic hollow cylinders with a wall thickness of 6-10 mm bonded to a metallic backing tube[6]. In the present work, the features of the sputtering process using those rotary targets and obtained film properties are discussed.

EXPERIMENTAL

ITO ceramic hollow cylindrical bodies were manufactured using the developed "ceramic" manufacturing process starting from in-house produced In_2O_3 powders and without pressure assistance during firing cycle[6]. ITO ceramics with an oxide ratio In_2O_3/SnO_2 of 90/10 was used in the present study (ITO ceramics with In_2O_3/SnO_2 ratio of (98-80)/(2-20) are generally employed in accordance with the customers' requirements). The finished ITO ceramic hollow cylinders were bonded to a metallic (Ti) backing tube using indium solder as a bonding material. The assembled and bonded sputtering targets may have a total length of the ceramic part up to 3.8 m. In the present work, the targets of 1.2-m length consisting of hollow cylinders with inside diameter of 133 mm and wall thickness of 6 mm (Fig. 1) were used for the sputtering campaign. All the ceramic segments had fired densities of 99% of TD or greater (density was determined using a water immersion method based on Archimedes law). The ceramic bodies after post-firing machining and the assembled targets after bonding were inspected using ultrasonic testing equipment, and no defects in ceramics and bonding layer were observed during inspection.

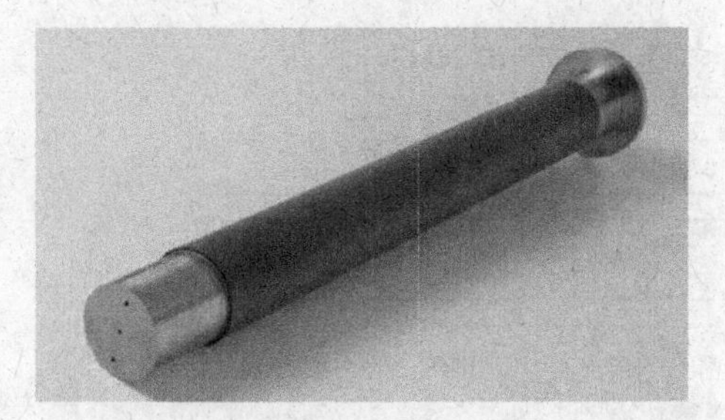

Fig. 1. ITO Ceramic Rotary Sputtering Targets (1.2-m Length of ITO Components)

Sputtering was conducted in collaboration with a distinguished sputtering system manufacturer using a conventional DC magnetron sputtering system providing rotation of the target. The target was tested at different sputtering power values starting from 5kW up to 12 kW until a clear racetrack was formed at both ends of the target (at that point, less than 1 mm of ITO was left in the racetrack closing). A total energy at sputtering was greater than 2,500 kW-hr. Sputtering behavior (e.g. plasma behavior, cycling performance, film uniformity) was evaluated. The targets were exposed to cyclic sputtering load (several thousands cycling) with the cycles of plasma ON mode during 25 s. and then 45 s. in plasma OFF mode. Arc formation was controlled using a power supply with an arc suppression unit. The sputtering process was conducted in Ar atmosphere with addition of O_2 reactive gas in various quantities. The coatings were deposited at ambient temperature and onto preheated (235°C) glass substrates (Corning) without using Plasma Emission control[7] for the reactive gas flow; also some coating samples of the as-deposited films were annealed at 200°C during 60 min.

Microstructure and phase composition of the films were studied using SEM and XRD techniques, respectively. Film thickness was optically determined using a reflectometer analyzer (J.Y. Horiba, Tokyo, Japan). Specific electrical resistivity of the films was measured using a four-point probe measuring unit (Jandel Scientific, Corte Madera, CA). Transmittance in the visible range from 400 to 800 nm wavelength was measured using a spectrophotometer (Perkin-Elmer, Norwalk, CT). Film stress was determined for selected film samples deposited onto Si wafers of 100 mm diameter in accordance with a standard procedure for FPD applications using cantilever technique; an α-step profilometer was used to measure the bending contour due to the stress in the thin films.

RESULTS AND DISCUSSION

Sputtering Behavior

The erosion profile after the sputtering campaign of 2,400 kW-hr was smooth and uniform over practically the entire length of the target. It can be seen from Fig. 2 that relatively thick remaining areas are only near both ends of the target (about 50 mm length), and practically the entire target was spent with a thickness of the remaining material of 1.8-2.5 mm with a slight thicker area in the middle zone of the profile. Based on this diagram, rotary targets can be utilized for min. 80% for this specific magnetron, i.e. a few times greater than planar targets. The target utilization may be increased using a longer sputtering cycle; actually, the longer campaign allowed to reach the remaining target thickness of 1.3 mm, and about 90% of utilization could be attained. Also utilization may be increased by optimization of the magnetic field at the racetrack closing at both ends of the target. The use of a longer magnetron with a longer target will also promote the target utilization reducing the effect of the ends of the targets with lower utilization.

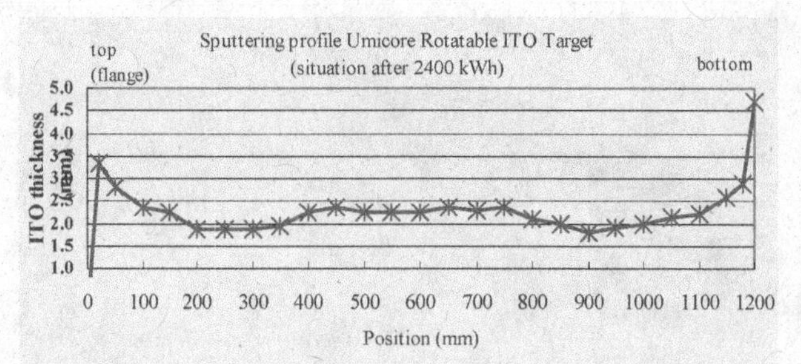

Fig. 2. Erosion Profile Measured on the Rotary Sputtering Target Surface after Total Sputtering Energy of 2,400 kW-hr

Sputtering was very stable with no changes in the target appearance and plasma ignition. No cracks were observed on the ceramic components of the target, arc traces were completely absent. Contrary to sputtering with ITO planar targets, nodules on the ceramic surface of the rotary targets due to re-deposition of sputtered material were not observed. As is generally known in ITO deposition practice, nodules ("black spots") may be observed at the edges of and outside the erosion race track of planar targets during long-cycle sputtering. The nodules are considered as indium sub-oxide In_2O[8, 9]. The nodules tend to cause electrical arcing and thin film deterioration, e.g. pinholes, and decrease of film uniformity. After some period of time, the target needs to be cleaned (i.e. nodules need to be removed mechanically). This not only reduces system up-time (the vacuum system needs to be vented and the cleaning procedure takes time), but the cleaning is a new source of particles in the deposition system, and the mechanical removal of nodules leads to loss of target material. In other words, the use of ITO rotary targets significantly increases process stability and reduces cost of ownership due to higher system up-time and improved product yield.

Basically, the sputtering power generates a considerable amount of heat that may affect thermal behavior of ITO ceramic bodies during sputtering. The ability to withstand thermal gradients during sputtering is defined by not only thermal shock resistance of ITO bodies that is generally not very high due to the nature of this ceramic, but also, in the valuable extent, by the sputtering cycle, cooling efficiency, rotation speed and some other factors. In order to provide the targets integrity, the sputtering cycle, including sputtering power ramp, was conducted very gentle. The quality of the bonding that defines cooling efficiency of the target during sputtering was adequate (in the case of bonding problems, cooling becomes not efficient due to inappropriate heat and cooling transfer at the local areas of the target, and this thermal mismatch can reach the sufficient values, which ITO ceramic bodies cannot withstand) that was inspected using non-destructive ultrasonic technique.

Thin Film Properties

ITO films obtained by deposition at room temperature are amorphous that was confirmed by XRD analysis, while thin films deposited onto preheated substrates have a polycrystalline structure with cubic shape In_2O_3 (ITO) grains (Fig. 3). Annealed films also have a polycrystalline structure with some preferential orientation. In both cases for the thermal activated deposition processes, the grains have sizes of 10-30 nm. The presence of other crystalline phases in the films was not detected. I.e. the thin film morphology and phase composition are identical to those films obtained from planar sputtering targets[5].

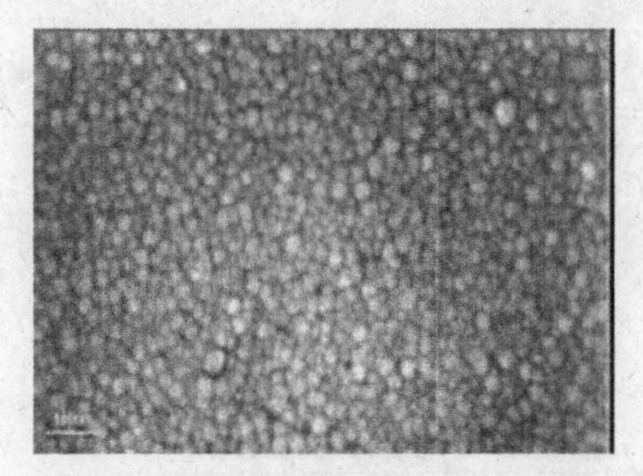

Fig. 3. Microstructure of ITO Film Obtained after Sputtering on the Preheated Substrate

The obtained thin films had a uniform thickness; the film uniformity was determined optically by measuring thickness in various points over the large area of a substrate. The thickness difference between different points of measuring was below 10% that is well accepted in the optoelectronic industry. Roughness of the films deposited at elevated temperature and after annealing was about 2 nm. The topography of the coatings is shown on Fig. 4.

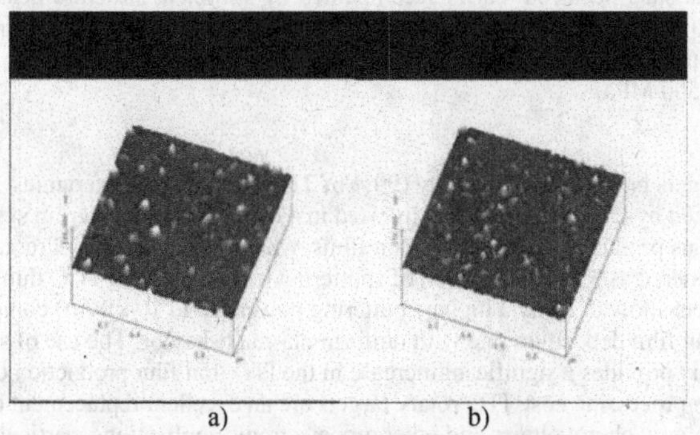

a) b)

Fig. 4. ITO Coating Topography (a- as-deposited, b- annealed film)

ITO thin films deposited using a rotary magnetron have properties (electrical resistivity and transmittance), which are well comparable with the properties, obtained with planar magnetrons. Specific electrical resistivity of 150-nm films sputtered onto heated substrates (at 235°C) is 180-200 μOhm.cm, i.e. these values are very acceptable for the FPD industry. Figure 5 demonstrates similarity of specific electrical resistivity values for the coatings deposited from both planar and rotatable magnetrons as a function of the composition of the Ar/O_2 reactive gas mixture. All the as-deposited ITO coatings (at elevated substrate temperature) with a resistivity below 2.E-04 Ohm.cm had optical transmittance at wavelength of 550 nm (with the glass substrate subtracted) of about 90-93 %, regardless of the deposition magnetron. The coatings after annealing at 200°C demonstrated even higher transmittance up to 93% due to formation of nanocrystalline structure.

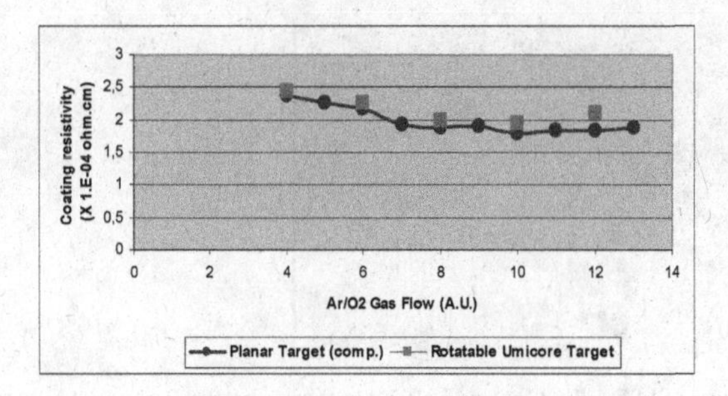

Fig. 5. ITO Film Resistivity for the Coatings Obtained from Ceramic Planar and Rotary Sputtering Targets

The film samples were evaluated for their stress condition. A value of about -195 MPa (i.e. compressive stress) was determined for as-deposited 150-nm films, while a value of about +250 MPa (tensile stress) was determined for thin films after annealing when the structure transformation from amorphous to polycrystalline state occurred that is accompanied by sufficient structure densification. These values are similar to the stress conditions determined for ITO films obtained using planar targets[5], and they are within the specification for the stress conditions established for the FPD applications (below than +/-500 MPa).

CONCLUSION

Rotary sputtering targets based on high-density (99% of TD or greater) ITO ceramics developed and commercialized by UIP were successfully used in a conventional sputtering system under stable plasma conditions producing high quality thin films, which satisfy industrial requirements. Such rotary targets are considered as a new generation of sputtering targets used for TCO thin film preparation. The rotary targets allow to apply a higher sputtering power (up to 10 kW/m) compared to planar equivalents, increasing film deposition rate with uniform plasma behavior. The use of sputtering targets with ceramic cylinders provides a significant increase in the ITO thin film production efficiency and overall reduction of film processing cost. ITO rotary targets are an excellent replacement of ITO planar targets in FPD technology, photovoltaics and other optoelectronic applications, particularly because of their significantly higher utilization of expensive ITO ceramics (about 80-90% of target utilization depending on the sputtering magnetron) and reduced re-deposition (i.e. absence of nodule). The comparison of sputtering efficiency using rotary and planar targets is summarized in Table.

Table. Rotary vs. Planar Magnetron Sputtering Technology

Processing features	Planar	Rotary
Target utilization (weight of target material sputtered vs. initial weight)	Low (25-40%)	High (80-90%)
Influence of magnet array configuration on target utilization	Strong	Weak (quasi absent for "dog-bone" shape targets)
Stability of sputtering process	Medium	Very high
Max. power density on target	Medium	High
Nodule formation	Possible	Quasi absent
Particle generation on target in process	Medium	Very low (quasi absent re-deposition)
System up-time	Low	Very high
Substrate radiation heating due to target	Medium	Low

REFERENCES

[1]J.L. Vossen, "Transparent Conducting Films", *Phys. Thin Films*, Ed. By G. Haas, M.H. Francombe, and R.W. Hoffman (Academic, New York) **9**, 1-71 (1977)
[2]D.S. Ginley, C. Bright, "Transparent Conducting Oxides", *MRS Bull.*, **8**, 15-18 (2000)
[3]T. Minami, "New n-Type Transparent Conducting Oxides", *MRS Bull.*, **8**, 38-44 (2000)
[4]I. Hamberg, C.G. Granquist, "Evaporated Sn-Doped In_2O_3 Films: Basic Optical Properties and Applications to Energy-Efficient Windows"; *J. Appl. Phys.*, **60**, R123-160 (1986)
[5]E. Medvedovski, N.A. Alvarez, O. Yankov, M.K. Olsson, "Advanced Indium-Tin Oxide Ceramics for Sputtering Targets"; *Ceramics International,* **34**, 1173-1182 (2007)
[6]E. Medvedovski, C.J. Szepesi, O.Yankov, M.K. Olsson, "Rotary ITO Ceramic Sputtering Targets for Transparent Conductive Thin Films Coating", *Amer. Ceram. Soc. Bull.*, 87, N. 2, 39-42 (2008)
[7]S. Schiller, U. Heisic, C. Korndorfer, et al., "Reactive DC High-Rate Sputtering as Production Technology", *Surface and Coatings Technology*, **33**, 405-423 (1987)
[8] S. Ishibashi, Y. Higuchi, Y. Oka, et al, "Low Resistivity Indium-Tin Oxide Transparent Conductive Films. II. Effect of Sputtering Voltage on Electrical property of Films", *J. Vac. Sci. Technol. A*, **8** (3), May/June, 1403-1406 (1990)
[9] P. Lippens, A. Segers, J. Haemers, et al., "Chemical Instability of the Target Surface during DC-Magnetron Sputtering of ITO-Coatings", *Thin Solid Films*, **317**, 405-408 (1998)

ULTRASONIC NON-DESTRUCTIVE TESTING OF CERAMIC SPUTTERING TARGETS

Eugene Medvedovski, Christopher J. Szepesi
Umicore Indium Products
50 Sims Ave., Providence, RI 02909, USA

ABSTRACT

Detection of defects during manufacturing of ceramic sputtering targets is important in order to prevent problems in the sputtering process and to produce high quality thin films, as required by the industry. This is especially important for large size targets with planar and, moreover, rotary configurations. Defects in bare ceramic components, such as forming flaws, cracks, and voids, and in bonded sputtering targets, such as de-bonds between the ceramic and metallic backing, have to be eliminated. Ultrasound C-scan systems were utilized for non-destructive inspection of indium tin oxide (ITO) ceramic sputtering targets. The testing was conducted for planar and rotary ITO ceramic components with densities up to 99.5% of TD and for actual bonded targets. Instruments with various designs were used for the testing, including systems, which were capable of inspecting long rotary targets with a total length up to 3 m or more. Transducers with frequencies of 10-30 MHz were used for ultrasonic evaluation. Microscopy studies and dye penetration testing of ceramic components were also used to confirm the presence of defects found by ultrasound C-scanning, and to understand their origin. Non-destructive ultrasonic evaluation can be successfully used as a quality control in the manufacturing of ceramic sputtering targets.

INTRODUCTION

The quality of ceramic sputtering targets affects the quality of the transparent conductive thin films, which are widely used as electrode layers in optoelectronic devices, as components of solar cells, and other applications. Ceramic sputtering targets used in industry for optoelectronic and photovoltaic applications usually consist of several ceramic components (tiles or hollow cylinders) bonded to metallic backing materials[1,2]. The films are commonly deposited by conventional DC magnetron sputtering or by some other techniques onto glass or polymer substrates. The film manufacturing requires a fine-tune deposition process and high quality sputtering targets. Defects in the ceramic components of the sputtering target, such as forming flaws, microcracks and voids, as well as voids in the bonding layers between the ceramic and backing materials of the targets, must be eliminated to attain high quality homogeneous films, to prevent problems in sputtering, and to maintain film processing efficiency with a minimum of downtime. For example, microcracks or voids in the ceramic components can result in the formation of particulates or nodules, which significantly reduce the film uniformity. Particle redeposition and nodule formation require downtime in sputtering and cleaning of the targets. The presence of defects in the bonding layer (e.g. de-bond spots when ceramic or metallic backing components have low adhesion with the solder) results in problems with heat transfer in the sputtering targets during film processing that causes cracking of the target, unpredicted downtime, and lower target utilization. The absence of defects in the targets is especially important for large size targets with planar and, moreover, rotary configurations, which have recently been employed in the industry[1,2].

Non-destructive testing (NDT) can be utilized for the identification of defects in the sputtering target components in order to prevent problems during and after processing. NDT may help not only to prevent the usage of the components with defects, but also may be useful for analyzing defect location, size, and distribution; this analysis is instrumental in the investigation of manufacturing problems, route cause analysis, and process improvement. There are a number of NDT methods used in industry. They include dye penetration testing and radiography, methods based on the measurement of electrical current, ultrasound and heat propagation, and some others. The features of these methods, their utilization, and positive and negatives points are summarized in Tables 1 and 2. Based on a comparison of these methods with the particular features of sputtering targets design and materials involved, ultrasonic C-scan systems were selected and utilized as the NDT method.

The ultrasound C-scan imaging technique is based on the detection of acoustic waves in different materials, which may be sensitive to differences in microstructure and phase composition. An initial electrical pulse is sent to the piezoelectric crystal in an ultrasonic transducer, and this pulse is converted into sound waves. These waves are sent to the inspected material, and their energy may be transmitted, reflected or scattered depending on the difference in acoustic impedance of the studied materials or the phases present in these materials. Any feature within the sample microstructure that changes the sonic velocity, such as the back surface of the sample, a crack surface, or an interface between two dissimilar materials, will reflect an echo back to the transducer/detector. Reflected signals are converted back into an electrical pulse and displayed in different amplitudes in an X-Y coordinate system. The collected pulse amplitude data are assigned to a color scale that indicates changes in the amplitude over the inspected area, which are displayed as a C-scan image.

Raw data are collected as sonic amplitude over a period of time comparable to the frequency of the ultrasonic waves. If the velocity of sound within the inspected material(s) is known, then the time period observed in the raw data can be converted to specific depth below the sample surface. Electronic gates can be set within the data collection software to create C-scan images from data at specific depths or certain thicknesses within the sample.

Ultrasonic transducers can be fabricated to operate at a variety of frequencies, typically between 1 and 30 MHz. The detectable flaw size is directly proportional to the ultrasound frequency. Depth of penetration, however, is inversely proportional to the ultrasound frequency and directly proportional to the density of the sample. For example, ultrasonic inspection of microcircuitry may employ a transducer that operates at a frequency of greater than 100 MHz, while a steel ingot may be inspected with a 5 MHz transducer. The operating frequency must be optimized for a particular material and flaw size. The ultrasound waves also interact more efficiently with features perpendicular to their direction of propagation, so transducers may be positioned at different angles to the sample surface in order to detect flaws oriented in various directions.

EXPERIMENTAL

The studies and actual NDT inspections were conducted for indium tin oxide (ITO) ceramic sputtering targets with planar and rotary designs, which are commercially manufactured at Umicore Indium Products (UIP). The ITO ceramic manufacturing and physical properties, as well as the target designs, have been described elsewhere[1,2]. ITO components (tiles and hollow cylinders of various sizes) have densities of 99% of TD or greater, and the ceramics typically display a very low electrical resistivity. Titanium was selected as the backing material for the assembled target, while indium alloys were used as the bonding material.

C-scan instruments with various designs were used for the testing, including systems capable of inspecting planar and rotary targets with a total length of 3 m or more. The inspected bodies were immersed into a tank with water as the ultrasound conduction medium. Transducers with frequencies of 10-30 MHz, positioned perpendicular to the sample surface, were used for ultrasonic evaluation. Microscopy studies (scanning electron microscopy SEM and optical microscopy) and dye penetration testing of ITO ceramic components were also used to confirm the presence of defects found by ultrasound C-scanning, and to understand their origin.

RESULTS AND DISCUSSION

Samples were inspected in as-fired, post-ground, or post-bond states. Initially, samples were chosen for inspection with visible surface flaws; several lots of samples were also inspected as a quality control measure regardless of the presence of visible flaws. As a complement to the ultrasonic inspection, cylindrical ITO ceramics were also examined with a fluorescent dye penetration technique.

Examples of C-scan images for two as-fired, rectangular ITO ceramic tiles are given in Figure 1. Two images were generated with different ultrasound inspection instruments operating at 20 MHz. Three surface cracks, surface roughness, and some internal porosity are visible. The differences in color are a result of the software settings. It should be noted that the surface cracks observed in this sample were visible to the naked eye, but the internal porosity was not.

To illustrate the effect of transducer frequency on image resolution, a ground ITO cylinder was inspected with 10 MHz and 20 MHz ultrasound waves. Figure 2 illustrates the same defects detected with both frequencies; however, image quality and resolution are greater when the transducer with thee higher frequency was used. Transducers with frequency of 30 MHz provide even higher image quality and resolution, especially for the ceramic components with sufficient thickness.

The origin of detected flaws was also investigated with microscopy techniques. For example, several cracked samples were sectioned and observed in an SEM to examine the microstructure of the cracked area. As shown in Figure 3, cracks did not extend through the entire sample cross-section. In addition, crack surfaces known to have resulted from mechanical fracture appeared rougher than the crack surfaces detected with ultrasound. It was subsequently concluded that the observed cracking did not occur as a result of mechanical stress, but may have been a result of thermal stress.

Figure 4 illustrates an example of sub-surface crack features, which are detectable with ultrasound but not with other surface-sensitive techniques, such as dye penetration. In this sample, two cracks begin at the sample's edge and extend at an angle to the inner/outer surfaces (instead of perpendicular, as typically seen). With the aid of the ultrasonic C-scan image, the length of each crack can be identified, and the entire cracked portion can be removed to salvage a part of the sample for its end use.

Since ultrasonic imaging is more sensitive to features perpendicular to the scanned surface, this technique is ideal for inspecting the degree of bonding between a ceramic component of a sputtering target and its backing support. Typically, ITO ceramic components are bonded to a metallic backing plate or tube with a metallic solder. This ensures mechanical stability of the target assembly and an acceptable thermal conductivity through the target, which is necessary to relieve thermal stresses during sputtering. Ultrasonic inspection can be used to detect areas of poor bonding, as illustrated in Figure 5, for rotary ITO sputtering target assemblies using a 20 MHz transducer. Un-bonded areas consist of an interface with air that reflects the sound waves more than the surrounding bonded areas. De-bonds, therefore, are detected and displayed in the C-scan image. The relative proportion of bonded and de-bonded areas can then be calculated using the imaging software. Any anomalous bonding features present in the inspected target can also be identified for quality control purposes.

The conducted studies were very helpful for the quality control of bare ceramic sputtering targets components and bonded targets; they allowed to exclude defective pieces from the manufacturing process and to supply the customers with the targets without defects. These studies were also helpful for the analysis of the manufacturing focused on the process improvement.

CONCLUSION

C-scan ultrasonic NDT technique was successfully used for inspection of ITO ceramic sputtering targets with planar and rotary configurations; both bare ceramics and actual bonded targets were tested. C-scans were collected and analyzed for the bodies over different areas. This method is capable to detect microcracks and flaws in the ceramic bodies and voids and bonding defects in the bonding area of the targets. Non-destructive ultrasonic evaluation can be successfully used as a routine quality control in the manufacturing of ceramic sputtering targets.

ACKNOWLEDGEMENT

The authors are grateful to Klaus Leitner (Umicore Materials AG, Balzers, Liechtenstein) for the helpful discussions. The authors are thankful to Applied Metrics, Inc., Fremont, CA; Matec, Inc., Northborough, MA; Physical Acoustics Corporation, Princeton Junction, NJ, and Sonoscan, Inc., Elk Grove Village, IL.

REFERENCES

[1]E. Medvedovski, N.A. Alvarez, O. Yankov, M.K. Olsson, "Advanced Indium-Tin Oxide Ceramics for Sputtering Targets"; *Ceramics International,* **34**, 1173-1182 (2007)

[2]E. Medvedovski, C.J. Szepesi, O.Yankov, M.K. Olsson, "Rotary ITO Ceramic Sputtering Targets for Transparent Conductive Thin Films Coating", *Amer. Ceram. Soc. Bull.*, 87, N. 2, 39-42 (2008)

Table 1. NDT Methods Comparison

Method	Characteristics detected	Advantages	Limitation
Dye penetration	Crack and other surface defects	Inexpensive, simple, easy to use and to understand, possible to automate	Only open surface may be examined. Not for porous materials or very rough surface
Ultrasonic	Change of acoustic sound or impedance due to cracks, voids, inclusions, interfaces, or defects in bonding	Good crack or void detection, usable for thick bodies, can be automate, computerized with recording and saving images	Requires coupling to material by contact to surface or by immersion to liquid. Not for very rough surfaces
X-ray	Change in density due to voids, inclusions, interfaces and materials variation	Used for various materials, thickness, shapes; versatile; data can be recorded	Serious radiation safety requirements, not easy to examine if cracks are perpendicular to X-ray direction. Very expensive
Electrical current	Changes in electrical conductivity due to materials variation, cracks, voids, inclusions	Possible to automate; relatively inexpensive	Only for rather conductive materials. Limited penetration depth
Magnetic	Leakage magnetic flux due to surface or near-surface cracks, voids, inclusions	Good sensitivity to the flaws; relatively inexpensive	Only for ferromagnetic materials. Limited penetration depth (used only for surface or near-surface defects); surface preparation may be required
Thermography	Changes in thermal conductivity due to materials variation, cracks, voids, inclusions	Possible to automate	Limited penetration. Not very high accuracy. Rather expensive

Table 2. NDT Methods Comparative Characteristics

Characteristics	Dye penetration	Ultrasonic	X-Ray	Electrical current	Magnetic	Thermography
Capital cost	Low	Medium-High	High	Low-Medium	Medium	High-Medium
Consumable cost	Low-Medium	Very low	High	Low	Medium	Low
Time of results	Immediate	Immediate	Delayed	Immediate	Short delay	Short delay
Geometry effect	Not very important	Important	Important	Important (thickness limit)	Not very important	Important
Access problems	Important	Important	Important	Important	Important	Important
Type of defect	Surface	Internal	Surface+internal	External mostly	External	Internal
Sensitivity	Low	High	Medium	High (not for all materials)	Low	Low-Medium
Formal record	Unusual	Expensive	Standard	Expensive	Unusual	Expensive
Operator skills required	Low	High	High	Medium	Low	High
Training required	Not important	Important	Important	Important	Important	Important
Portability of equipment	High	Medium	Low	Medium-High	Medium-High	Medium-High
Dependence on material composition	Little	High	Quite	High	High (magnetic only)	High
Ability to automate	Fair	Good	Fair	Good	Fair	Good
Capabilities	Defects only	Thickness gaging	Thickness gaging	Thickness gaging	Defects only	Thickness gaging

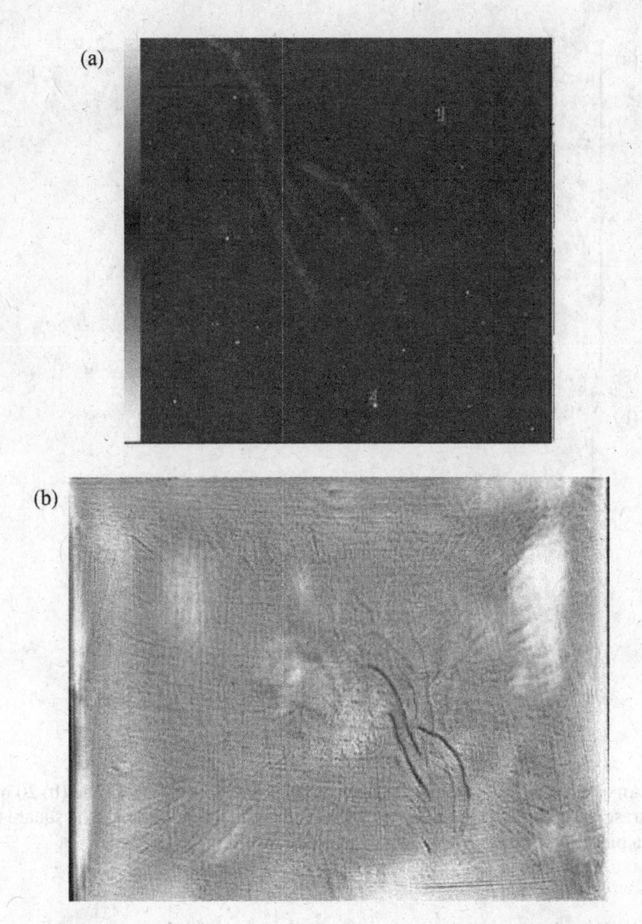

Figure 1. An as-fired, rectangular ITO ceramic tile inspected with two different ultrasound instruments. Three surface cracks can be identified in each image; image (a) displays internal porosity (white dots), while image (b) illustrates features associated with surface roughness.

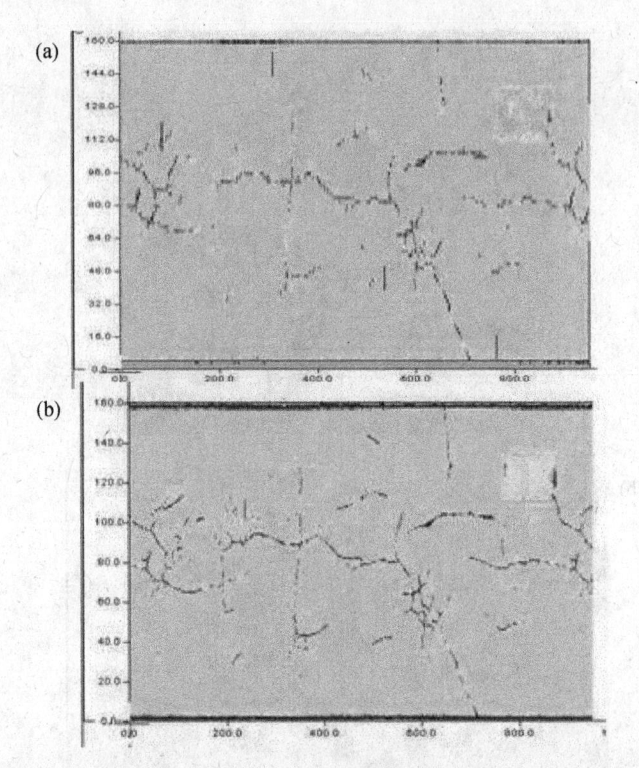

Figure 2. An as-ground ITO ceramic cylinder inspected with (a) 10 MHz and (b) 20 MHz ultrasound transducers to illustrate the effect on image resolution. The square feature is a piece of tape used for reference of position.

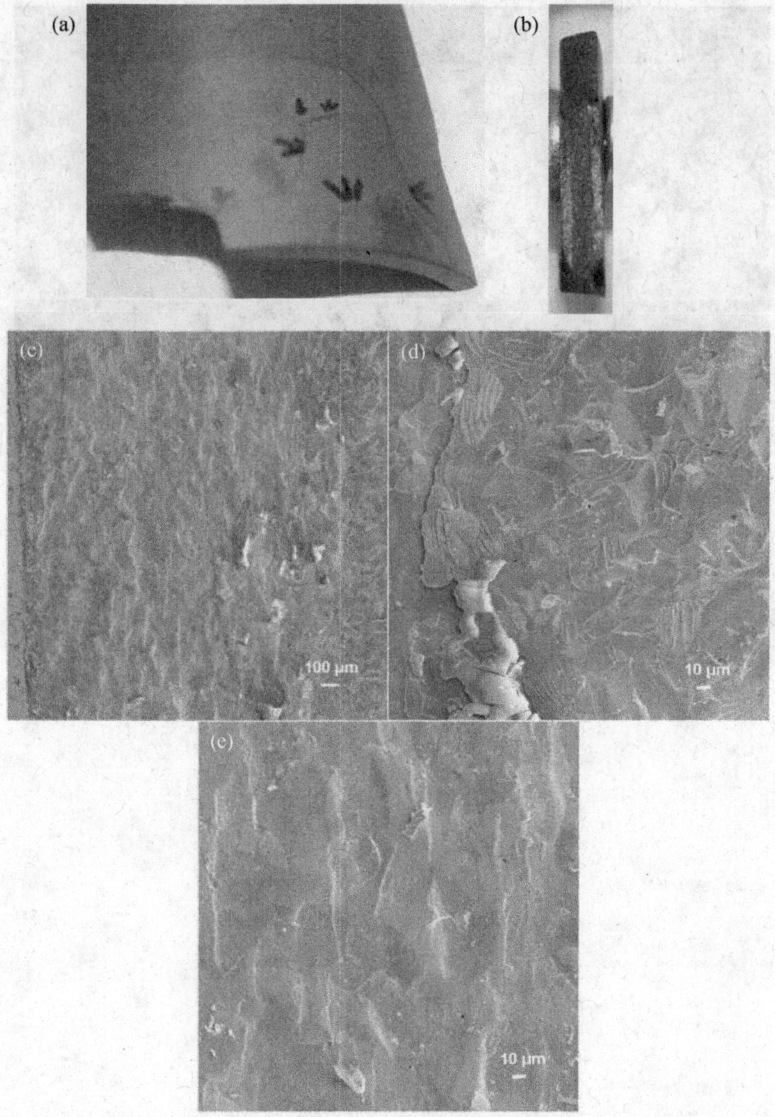

Figure 3. (a) Cracks detected with ultrasound in a ground ITO cylinder mechanically opened to view (b) crack surfaces. (c) SEM images illustrate a difference in crack surface roughness between (d) mechanical fracture surfaces and (e) previously-detected crack surface, suggesting that they have different origins.

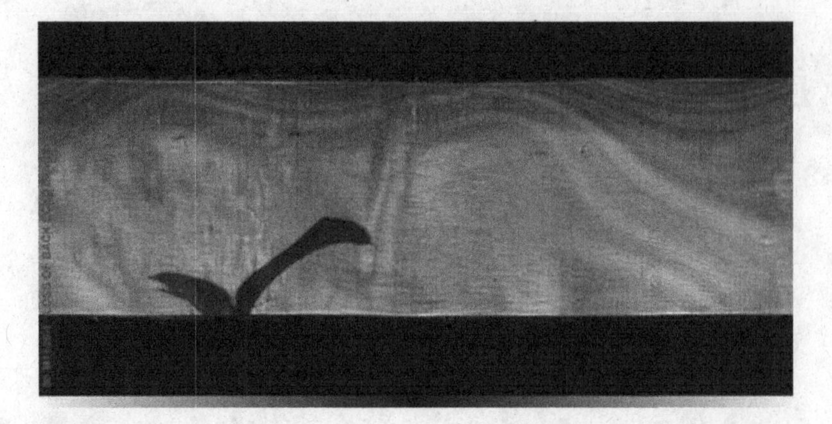

Figure 4. Cracks in this sample (black areas beginning at the bottom edge), which are not perpendicular to the sample surface. This is not apparent with other inspection techniques, such as dye penetration.

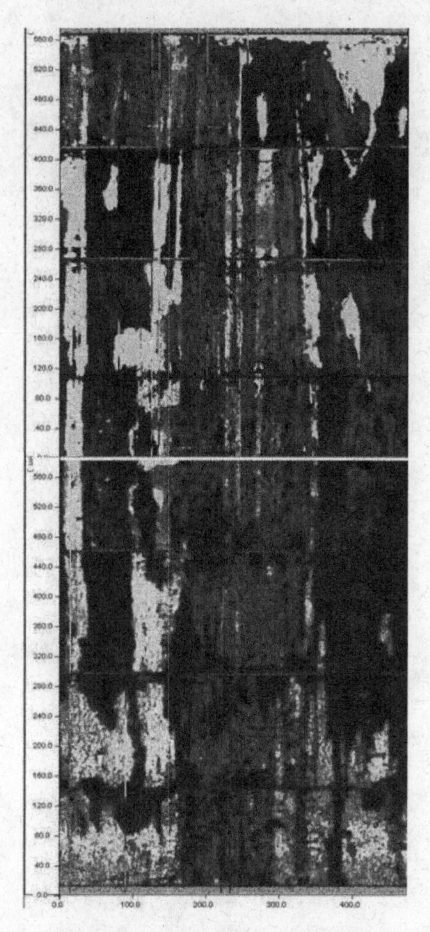

Figure 5. The quality of bonding between the ceramic and backing tube in a rotatable sputtering target assembly can be determined with ultrasonic NDT. Un-bonded areas contain an air interface that is detected and displayed on the C-scan image as lighter areas. Area of good bonding was calculated to be 83%.

MORPHOLOGY CONTROL OF METAL OXIDES FOR ENVIRONMENTAL SENSORS

Yoshitake Masuda[1], Xiulan Hu[1], Tatsuo Kimura[1], Kazumi Kato[1],
Masako Ajimi[2], Makoto Bekki[2] and Shuji Sonezaki[2]
1 National Institute of Advanced Industrial Science and Technology (AIST)
2266-98 Anagahora, Shimoshidami, Moriyama-ku, Nagoya 463-8560, Japan

2 TOTO Ltd. Research Laboratory, 1-1, Nakashima 2-chome, Kokurakita-ku, Kitakyushu, Fukuoka 802-8601, Japan

ABSTRACT

FTO (SnO_2: F) substrates were immersed in aqueous solutions to cover with TiO_2 films. They were assembly of nano-acicular crystals of anatase phase. The crystals were stood perpendicular to the substrates to grow along *c*-axis. Surface of the films had nano-sized relief structure. Dye adsorption property of the films was higher than that of poly-crystal films consisted of TiO_2 nano-particles (P25). The films were suitable for environmental sensors and dye-sensitized solar cells.

INTRODUCTION

Titanium dioxide (TiO_2) thin films are of interest for various applications including microelectronics[1], optical cells[2], solar energy conversion[3], highly efficient catalysts[4], microorganism photolysis[5], antifogging and self-cleaning coatings[6], gratings[7], gate oxides in MOSFETs (Metal-Oxide-Semiconductor Field Effect Transistor)[8, 9], etc. Accordingly, various attempts have been made to fabricate thin films and micropatterns of TiO_2, and in particular, the method based on an aqueous solution is important as an environment-friendly synthesis process, i.e., "green chemistry".

TiO_2 films on transparent conductive substrates such as FTO (Fluorine doped Tin Oxide) or ITO (Indium Tin Oxide) have been required for environmental sensors[10] and dye-sensitized solar cells[11, 12].

Surface area and surface chemical properties have much influence on dye adsorption property of the films. They are important factors for the sensors and solar cells.

In this study, we fabricated anatase TiO_2 films in aqueous solutions containing ammonium hexafluorotitanate ($[NH_4]_2TiF_6$) and boric acid (H_3BO_3)[13]. They were grown at 50 °C for several hours. Crystalline anatase TiO_2 films were formed on the substrates without annealing in this system. Dye adsorption property of the films was evaluated after annealing. It was 3 times higher than that of the sintered films consisted of nano-sized TiO_2 particles (P25). Additionally, adsorption property of the films without annealing was two times higher than the films with annealing[13].

EXPERIMENTAL

Glass substrates coated with F-doped SnO_2 transparent conductive films (FTO, SnO_2: F, Asahi Glass Co., Ltd., 9.3-9.7 Ω/□, 26×50×1.1 mm) were subjected to air-blowing to remove dust and were exposed to ultraviolet light (low-pressure mercury lamp PL16-110, air flow, 100 V, 200 W, SEN Lights Co., 14 mW/cm^2 for 184.9 nm at a distance of 10 mm from the lamp, 18 mW/cm^2 for 253.7 nm at a distance of 10 mm from the lamp) for 10 min. The initial FTO substrates showed a water contact angle of 96°. The UV-irradiated surfaces were, however, wetted completely (contact angle 0–1°). TiO_2 crystallization is sensitive to the substrate surface. OH groups were reported to accelerate crystallization of TiO_2 in the solution[14-18]. Super-hydrophilic OH surfaces were thus utilized to accelerate TiO_2 deposition in our study. Both ends

(26×14 mm) of the glass substrates were masked using Scotch tapes (CM-18, 3M) to prevent deposition[19].

Ammonium hexafluorotitanate ($[NH_4]_2TiF_6$) (Morita Chemical Industries Co., Ltd., FW: 197.95, purity 96.0%) and boric acid (H_3BO_3) (Kishida Chemical Co., Ltd., FW: 61.83, purity 99.5%) were used as received. Ammonium hexafluorotitanate (2.0096 g) and boric acid (1.86422 g) were separately dissolved in deionized water (100 mL) at 50°C[20]. Boric acid solutions (concentration 0.15 M) were added to ammonium hexafluorotitanate solutions (concentration 0.05 M). FTO substrates were immersed vertically in the middle of the solutions immediately after mixing of two solutions. The solutions were kept at 50°C for 48 h. Substrates were removed from the solutions after 2, 5, 25 and 48 h. The films were rinsed with distilled water and annealed at 500°C for 30 min in air after removal of the Scotch tapes. Dye adsorption properties of them were evaluated. TiO_2 films were annealed because of comparison with TiO_2 films consisted of nanoparticles (TiO_2 P25, Degussa). Additionally, dye adsorption property of the films immersed for 5 h was evaluated without annealing.

Morphologies of TiO_2 films were observed by field emission scanning electron microscope (FE-SEM; JSM-6335F, JEOL Ltd.) and transmission electron microscope (TEM; H-9000UHR, 300 kV, Hitachi). Crystal phases were evaluated by X-ray diffractometer (XRD; RINT-2100V, Rigaku) with CuKα radiation (40 kV, 30 mA). Diffraction patterns were evaluated using data from the ICSD (Inorganic Crystal Structure Database) (FIZ Karlsruhe, Germany and NIST, USA) and FindIt.

Polyvinyl chloride tape (PVC, CH_2-$CHCl)_n$, 26×22 mm, 100-μm thickness) was perforated with holes 25 mm in diameter using a flatbed cutting plotter (CG-60ST; Mimaki Engineering Co., Ltd.). The TiO_2 films were covered with PVC tapes.

ssDNA-Cy5 (cy5-DP53-t: Cy5-GCGGCATGAACCTGAGGCCCATCCT, dye labeling DNA) was dissolved in water. ssDNA-Cy5 solutions (100 nM) were dropped with pipettes onto the holes in the PVC tapes[10]. The films were dried at 95°C for 10 min in air. The films were then rinsed 3 times in sodium dodecyl sulfate (SDS, $NaC_{12}H_{25}SO_4$) for 15 min each time and rinsed 3 times in ultrapure water. SDS was used for cleaning of the substrates. Strongly-connected dyes were remained on the substrates and they contributed to photoluminescence property. They were then boiled in water for 2 min, immersed in dehydrated ethanol at 4°C for 1 min and dried by strong air flow.

Photoluminescence image and intensity were evaluated by Typhoon Trio scanner (GE Healthcare UK Ltd.) using excitation light of 633 nm (He/Ne laser)(Fig. 1).

Particulate films 1000 nm thick were formed using TiO_2 nanoparticles (TiO_2 P25, Degussa) and were sintered at 500°C for 30 min in air. Photoluminescence intensity of the particulate films and bare FTO substrates were evaluated after adsorption of 100 nM ssDNA-Cy5 for comparison.

RESULTS AND DISCUSSION

Deposition of anatase TiO_2 proceeds by the following mechanisms[20]:

$$TiF_6^{2-} + 2H_2O \rightleftarrows TiO_2 + 4H^+ + 6F^- \quad ...(a)$$

$$BO_3^{3-} + 4F^- + 6H^+ \longrightarrow BF_4^- + 3H_2O \quad ...(b)$$

Equation (a) is described in detail by the following two equations:

$$TiF_6^{2-} \xrightarrow{nOH^-} TiF_{6-n}(OH)_n^{2-} + nF^- \xrightarrow{(6-n)OH^-} Ti(OH)_6^{2-} + 6F^- \quad ...(c)$$

$$Ti(OH)_6^{2-} \longrightarrow TiO_2 + 2H_2O + 2OH^- \quad ...(d)$$

Fluorinated titanium complex ions gradually change into titanium hydroxide complex ions in the aqueous solution as shown in Eq. (c). Increase of F^- concentration displaces Eqs. (a) and (c) to the left; however, the produced F^- can be scavenged by H_3BO_3 (BO_3^{3-}) as shown in Eq. (b) to displace Eqs. (a) and (c) to the right. Anatase TiO_2 was formed from titanium hydroxide complex ions ($Ti(OH)_6^{2-}$) in Eq. (d).

The solution became cloudy 10 min after mixing the boric acid solution and ammonium hexafluorotitanate solution. The particles were homogeneously nucleated in the solution, making the solution white. They then gradually precipitated and covered the bottom of the vessel. The solution became slightly white after 5 h and transparent after 25 h. The solution changed from light white to transparent after roughly 10–15 h.

The substrates with films were dried in air after immersion. The films were colored slightly white[19]. Whiteness gradually increased as a function of deposition time due to increased film thickness. The films did not peel off during ultrasonic oscillation treatment in acetone for 30 min as they showed high adhesion strength.

Strong X-ray diffractions were observed for films deposited on FTO substrates and were assigned to SnO_2 of the FTO films[19]. Glass substrates without FTO coating were immersed in the solutions for comparison. Weak X-ray diffraction peaks were observed at 2θ = 25.3, 37.7, 48.0, 53.9, 55.1 and 62.7° for films deposited on glass substrates and were assigned to 101, 004, 200, 105, 211 and 204 diffraction peaks of anatase TiO_2 (ICSD No. 9852), respectively[19] (Fig. 2). A broad diffraction peak from the glass substrates was also observed at about 2θ = 25°. The 004 diffraction peak of anatase TiO_2 was not clearly distinguished for the films on FTO substrates because both the 004 diffraction peaks of TiO_2 and the strong diffraction peaks of FTO were observed at the same angle.

The average crystallite size (d) was estimated from the full-width half-maximum of the peaks in Fig. 2 with use of the Debye-Sherrer equation after peak separation and curve-fitting:
$d = k\lambda / \beta\cos\theta$

where λ is radiation wavelength (1.54051 Å), β is the peak halfwidth, θ is a diffracting angle, $k = 0.9$.

Crystallite size perpendicular to the (004) plane was estimated from the full-width half-maximum of the 004 peak to be 17 nm.

The films deposited for 48 h had assemblies of TiO_2 nanocrystals 5–20 nm in diameter on the surfaces (Fig. 3). Film thickness increased to 260, 360 and 600 nm at 2, 5 and 25 h, respectively, and reached 760 nm at 48 h [19]. The films were constructed of TiO_2 crystals and had uneven surface microstructures.

The liquid phase crystal deposition method has an advantage in that anatase TiO_2 is crystallized in the solutions without annealing. This would realize morphological control and formation of assemblies of TiO_2 nanocrystals. The solutions changed from milky white at 10 min to light white at 5 h and became transparent at 25 h. The solution condition would influence the deposition mechanism of TiO_2 films. TiO_2 particles homogeneously nucleated in the milky solutions deposit at the initial stage. TiO_2 nanocrystals observed on the surface would then crystallize in the transparent solutions after roughly 10–15 h.

TiO_2 films deposited for 25 h were observed by TEM after ultrasonication in water for 20 min. The films were constructed of two layers (Fig. 4). Under layers 200 nm thick were polycrystalline films of anatase TiO_2. Upper layers 300 nm thick were assemblies of acicular TiO_2 crystals that grew perpendicular to the substrates. The films were shown to be a single phase of anatase TiO_2 by electron diffraction patterns. Electron diffractions from the (004) planes were stronger than that of the (101), (200), (211) planes, etc. showing anisotropic crystal growth along the c-axis. Additionally, 004 diffractions were strong perpendicular to the substrates, showing that the c-axis orientation of acicular crystals was perpendicular to the substrates. Acicular TiO_2 crystals had a long shape, ~ 300 nm in length and 10–100 nm in diameter (Fig. 4). Anisotropic crystal growth and c-axis orientation were distinctive features of this system.

Strong visible luminescence from ssDNA-Cy5 was observed at holes. Photoluminescence brightness increased film thickness (Fig. 5). It was much higher than that from FTO substrate.

Increase of photoluminescence indicates that the adsorption amount of ssDNA-Cy5 increased with film thickness. This would be related to surface morphology and film thickness. It is particularly worth noting that photoluminescence intensity from the 760-nm-thick TiO_2 films was 3 times larger than that from the 1000-nm-thick particulate films constructed of nano TiO_2 particles (P25). Assembly of acicular TiO_2 crystals would increase the surface area and effectively adsorb ssDNA-Cy5 due to its nano/micro relief structure. BET surface area of P25 particles was reported to 50 m^2/g by Degussa. TiO_2 particles prepared by our method had BET surface area of 168-270 m^2/g[21-23]. Therefore, it was natural that photoluminescence intensity from the TiO_2 films prepared in this study was 3 times higher than that of thicker particulate films constructed of TiO_2 nanoparticles (P25).

It is also important to remember that luminescence intensity from the 760-nm films is much higher than that from the 260-nm films, the 360-nm films and the 600-nm films. This would be strongly related to the film surface morphology. The films deposited for 25 h were constructed of a dense layer and assembly of acicular crystals (Fig. 4). Dense polycrystalline film would be formed at the initial stage of immersion for roughly 10–15 h. Acicular TiO_2 was then crystallized to form an assembly on the film surface after roughly 10–15 h. The films deposited for 25 h were thus constructed of two layers. Dense polycrystalline films had low surface area and flat surfaces. Only a small amount of ssDNA-Cy5 was adsorbed on dense films deposited for 2 or 5 h. The film deposited for 25 h had an assembly of acicular TiO_2 that adsorbed a larger amount of ssDNA-Cy5 compared to the films deposited for 2 or 5 h. Acicular crystals would continue to grow on the surface of the films to increase film thickness. The films deposited for 48 h thus have assemblies of acicular crystals that grew well in the solutions. The films would have a high surface area and nano/micro relief structures suitable for adsorption of ssDNA-Cy5. Consequently, photoluminescence intensity improved dramatically from films deposited for 25 h to films deposited for 48 h.

Furthermore, annealing effect on dye adsorption property was evaluated. The FTO substrates were immersed for 5 h. One TiO_2 films were annealed at 500°C for 30 min in air as mentioned above. The other film was evaluated without high temperature annealing. Dye adsorption property of the film without annealing was two times higher than the film with annealing. It indicated that the TiO_2 films were well crystallized in the solutions and had high surface area. They were slightly sintered to decrease surface area by the annealing. Surface conditions and functional groups were also changed by the annealing. TiO_2 films without annealing are suitable for sensors which require high dye adsorption properties.

CONCLUSION

Assemblies of acicular TiO_2 crystals were formed on FTO substrates in aqueous solutions. The crystals grew along *c*-axis to form anisotropic shape. The films had high surface area due to nano-sized surface relief structure. Dye adsorption property of annealed films increased as film thickness. It reached to 3 times larger than that of annealed particulate films consisted of nano-sized TiO_2 particles (P25). Additionally, dye adsorption property of as-deposited films was 2 times larger than that of annealed films. TiO_2 films consisted of assemblies of acicular crystals contribute to development of next-generation devices such as environmental sensor to detect environmental toxin. The films have also high potential for dye-sensitized solar cells that require high surface area, high transparency and high electrical conductivity. The process developed here has other advantages that anatase nano-structures can be formed in aqueous solutions without high temperature annealing. Thus, the nano-structures of anatase TiO_2 can be formed on any kinds of materials and any shapes of substrates. Low heat resistant polymer films, nano-particle, complicated shaped substrates, for instance, can be easily coated with anatase TiO_2 nano-structures. Formation process of TiO_2 nano-structures can be also combined with bio-processes which should be operated at ordinary temperature in aqueous solutions. This technique will open next door for hybrid materials consisted of ceramics, polymers and bio-materials.

ACKNOWLEDGMENT

This work was supported by METI, Japan, as part of R&D for High Sensitivity Environment Sensor Components.

REFERENCES

[1] G.P. Burns, J. Appl. Phys. 65 (1989) 2095.
[2] B.E. Yoldas, T.W. O'Keeffe, Appl. Opt. 18/18 (1979) 3133.
[3] M.A. Butler, D.S. Ginley, J. Mat. Sci. 15/1 (1980) 1.
[4] T. Carlson, G.L. Giffin, J. Phys. Chem. 90/22 (1986) 5896.
[5] T. Matsunaga, R. Tomoda, T. Nakajima, N. Nakamura, T. Komine, Appl. Environ. Microbiol. 54/6 (1988) 1330.
[6] R. Wang, K. Hashimoto, A. Fujishima, Nature 388/6641 (1997) 431.
[7] S.I. Borenstain, U. Arad, I. Lyubina, A. Segal, Y. Warschawer, Thin Solid Films 75/17 (1999) 2659.
[8] P.S. Peercy, Nature 406/6799 (2000) 1023.
[9] D.J. Wang, Y. Masuda, W.S. Seo, K. Koumoto, Key Eng. Mater. 214/2 (2002) 163.
[10] H. Tokudome, Y. Yamada, S. Sonezaki, H. Ishikawa, M. Bekki, K. Kanehira, M. Miyauchi, Appl. Phys. Lett. 87/21 (2005) 213901.
[11] M.K. Nazeeruddin, F. De Angelis, S. Fantacci, A. Selloni, G. Viscardi, P. Liska, S. Ito, B. Takeru, M.G. Gratzel, J. Am. Chem. Soc. 127/48 (2005) 16835.
[12] P. Wang, S.M. Zakeeruddin, J.E. Moser, R. Humphry-Baker, P. Comte, V. Aranyos, A. Hagfeldt, M.K. Nazeeruddin, M. Gratzel, Adv. Mater. 16/20 (2004) 1806.
[13] Y. Masuda, M. Bekki, S. Sonezaki, T. Ohji, K. Kato, Thin Solid Films submitted.
[14] Y. Masuda, N. Saito, R. Hoffmann, M.R. De Guire, K. Koumoto, Sci. Tech. Adv. Mater. 4 (2003) 461.
[15] T.P. Niesen, M.R. DeGuire, J. Electroceramics 6/3 (2001) 169.
[16] R.J. Collins, H. Shin, M.R. De Guire, A.H. Heuer, C.N. Sukenik, Appl. Phys. Lett. 69/6 (1996) 860.
[17] H. Shin, R.J. Collins, M.R. De Guire, A.H. Heuer, C.N. Sukenik, J. Mater. Res. 10/3 (1995) 692.
[18] Y.F. Gao, K. Koumoto, Cryst. Growth Des. 5/5 (2005) 1983.
[19] Y. Masuda, K. Kato, Thin Solid Films 516 (2008) 2474.
[20] Y. Masuda, T. Sugiyama, W.S. Seo, K. Koumoto, Chem. Mater. 15/12 (2003) 2469.
[21] Y. Masuda, K. Kato, J. Jpn. Soc. Powder Powder Metallurgy 54/12 (2007) 824.
[22] Y. Masuda, K. Kato, Cryst. Growth Des. 8/9 (2008) 3213.
[23] Y. Masuda, K. Kato, Japan Patent No. Japanese Patent Application Number: P 2007-2402362007.

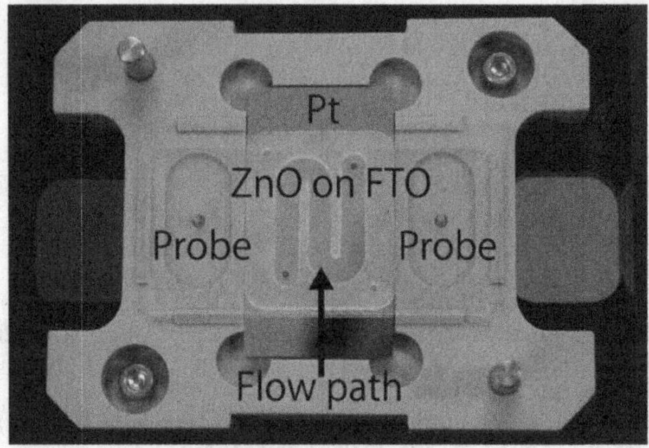

Fig. 1 Photograph of photoelectrochemical measuring cell.

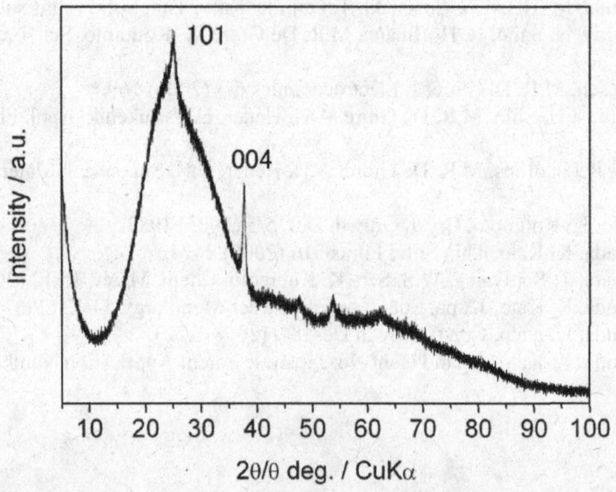

Fig. 2 XRD diffraction pattern of anatase TiO_2 film on glass substrate.

Fig. 3 SEM micrograph of anatase TiO_2 film deposited for 48 h.

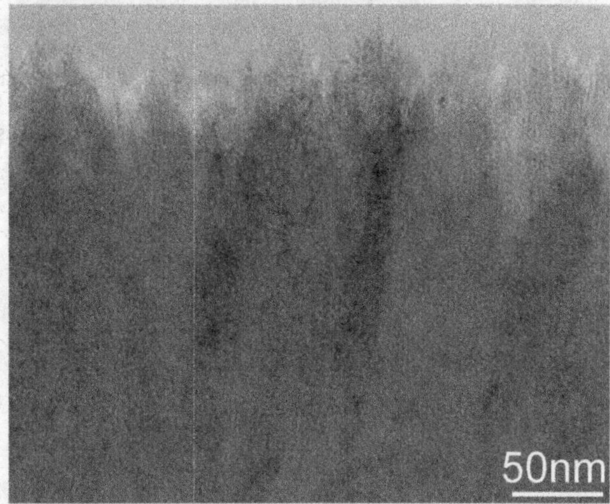

Fig. 4 Cross-sectional TEM image of acicular TiO_2 crystals.

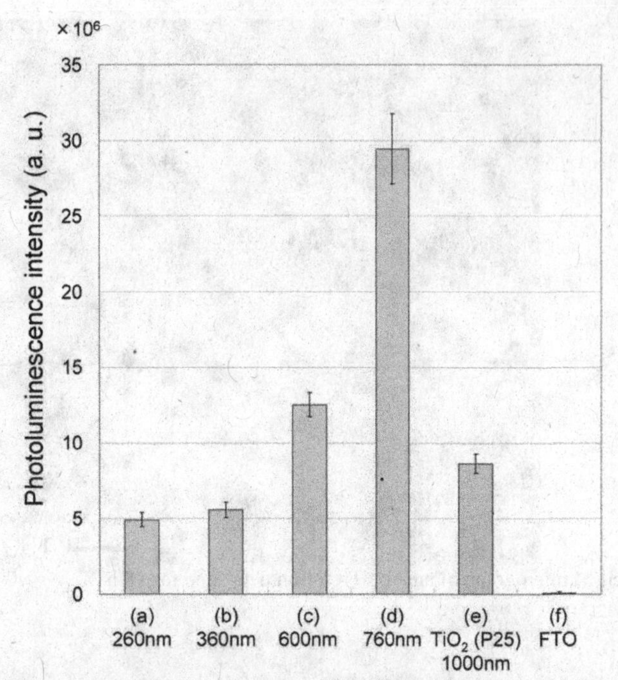

Fig. 5 Photoluminescence intensity of anatase TiO_2 film with varying thickness deposited by LPCD ((a)-(d)), 1000-nm-thick particulate film constructed of TiO_2 nanoparticles (e) and bare FTO substrate (f). All substrates were covered with ssDNA-Cy5 solution of 100 nM.

BASIC STUDY OF JOINT INTERFACE FORMATION IN MAGNETIC PRESSURE SEAM WELDING

Hisashi Serizawa, Isao Shibahara, Sherif Rashed and Hidekazu Murakawa
Joining and Welding Research Institute, Osaka University
11-1 Mihogaoka, Ibaraki, Osaka 567-0047, Japan

Mitsuhiro Watanabe
Materials and Structures Laboratory, Tokyo Institute of Technology
4259 Nagatsuta-cho, Midori-ku, Yokohama, 226-8502, Japan

Shinji Kumai
Department of Materials Science and Engineering, Tokyo Institute of Technology
4259 Nagatsuta-cho, Midori-ku, Yokohama, 226-8502, Japan

ABSTRACT

The magnetic pressure seam welding is one of the candidate methods to join thin sheet multifunctional materials. In this research, to examine the mechanism of magnetic pressure welding from a dynamic viewpoint, numerical simulation of the impact was carried out by using a commercial Euler-Lagrange coupling software MSC.Dytran (MSC.Software) as a first step of the computational studies, where the joint between Fe and Al was employed according to the previous experimental researches. From the serial numerical results, it was found that the increase of temperature at the joint interface was not enough to melt Al or Fe in the range of collision velocity and angle studied in this report. Also, it was revealed that the very large mean stress occurred at the interface which could be considered as the pressure at joint interface and Al moved with high velocity along the interface. Moreover, it was found that there were_two patterns of plastic strain distribution near the joint interface depending on the collision velocity and collision angle. Finally, it can be concluded that the plastic strain pattern might be related to the success of magnetic pressure seam welding.

INTRODUCTION

There have been strong demands to join multifunctional materials to other materials without any functional defects in the multifunctional materials for propagating their use. The ordinary physical joining methods using heat sources such as arc welding, laser welding and electron beam welding have to cause the microstructural changes at the joint interface [1,2] and might largely affect the original feature of the multifunctional materials. In order to overcome this problem, several joining methods have been proposed, which are, for examples, diffusion bonding [3], explosive bonding [4], friction welding [5], magnetic pulse welding and so on.

Recently, Aizawa et al. developed a seam welding technique called "magnetic pressure seam welding" as one of the species of the magnetic pulse welding [6]. In this joining method, a thin plate of a material with a high electric conductivity is suddenly subjected to a high density magnetic field and magnetic forces cause the plate to fly and impact a parent plate. Then, a seam is created between the two plates due to this impact. Although the large impact is applied at the interface between dissimilar materials, the temperature of the joint is close to room temperature and any superior material properties of these dissimilar materials seem to be preserved. So, the magnetic pressure seam welding is considered as one of the most candidate methods to join dissimilar thin sheet materials between the multifunctional materials and other materials.

Experimental investigations of the conditions necessary for successful joining, microstructural observations of the joint interface and mechanical evaluation of the joints were reported in the previous studies [7-10]. However, the mechanism of this joining process is still not well understood and then appropriate joining conditions have been decided from huge experimental struggles. The final target of this research is to examine the mechanism of the magnetic pressure seam welding from a dynamic viewpoint and to reveal the appropriate joining conditions theoretically. So, in this report, numerical simulation of the impact was carried out by using a commercial Euler-Lagrange coupling software MSC.Dytran (MSC.Software) [11] as a first step of the computational examination of the magnetic pressure seam welding.

MODELING FOR ANALYSIS

Magnetic Pressure Seam Welding

Figure 1 shows the principle of the magnetic pressure seam welding. An electrical discharge circuit is applied to the present welding. The circuit consists of a power supply, a capacitor, a discharge gap switch and a one-turn flat coil. A plate called "flyer plate" is set over the coil. A thin film is inserted between them as an insulator. Another plate called "parent plate" is placed so that it overlaps the flyer plate with a little gap. The parent plate is fixed firmly using a fixture. When an impulse current from the bank passes through the coil, a high density magnetic flux is suddenly generated around the coil. The generated high density magnetic flux lines intersect the overlapped area of the plates. Eddy currents are induced in this area, in particular, in a very surface layer of the flyer plate. Eddy currents flow in an opposite direction to the impulse current in the coil. The high density magnetic flux and the generated eddy currents induce an electro-magnetic force upward. This force drives a part of the flyer plate to the parent plate with an extremely high speed (150 - 500 m/s) [12].

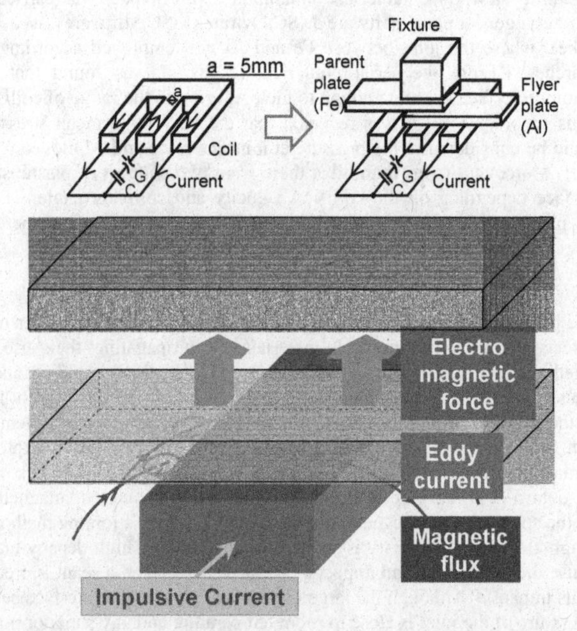

Fig. 1 Schematic illustration of magnetic pressure seam welding.

Whole Model for Magnetic Pressure Seam Welding

In the previous studies of the joint between Fe and Al using this joining process, the metal jet, whose composition was mainly Al, was observed [12]. So, Fe and Al were modeled by a Lagrange and an Euler model, respectively. In order to examine collision behavior in this process roughly, the whole plates of Fe and Al (200^L x 1^T mm) were simulated as two dimensional plain strain problem as shown in Fig. 2. Where a minimum element size was 100 x 100 μm^2 and total number of elements and nodes was 13240 and 29172, respectively. An initial gap between the plates was set to 1 mm and an initial velocity was applied to the center part of the flayer plate (Al), whose length was assumed to be 10 mm. Table 1 shows mechanical and thermal properties used in this research. According to the experimental results [12], initial velocity in this computation was varied in the range from 100 to 500 m/s. Figure 3 shows typical computational results of the collision behavior of the whole plates of Fe and Al, where the initial velocity was 200 m/s and the distributions of mean stress were represented. As shown in Fig. 3(b), Al plate collided with Fe at the center of the plate and a very large mean stress occurred. Also, after further 0.5 μs, the collision point moved along the surface of Fe plate. These behaviors have good agreements with the experimental results recorded by a high speed camera [10,12]. Moreover, from these computational results for the whole plates, it was found that the collision angle at the contact point between Al and Fe plate was monotonically increased after the first collision.

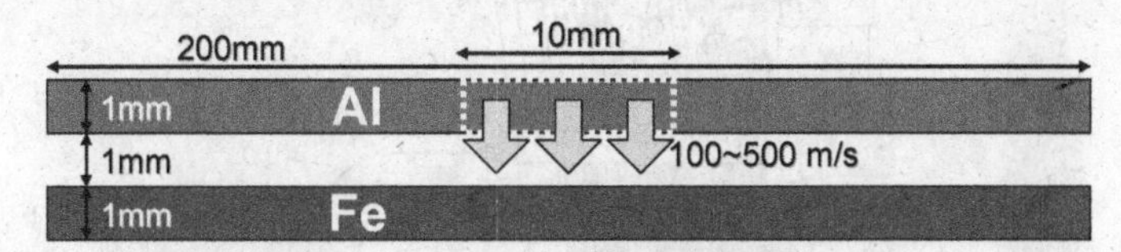

Fig. 2 Schematic illustration of whole model for magnetic pressure seam welding.

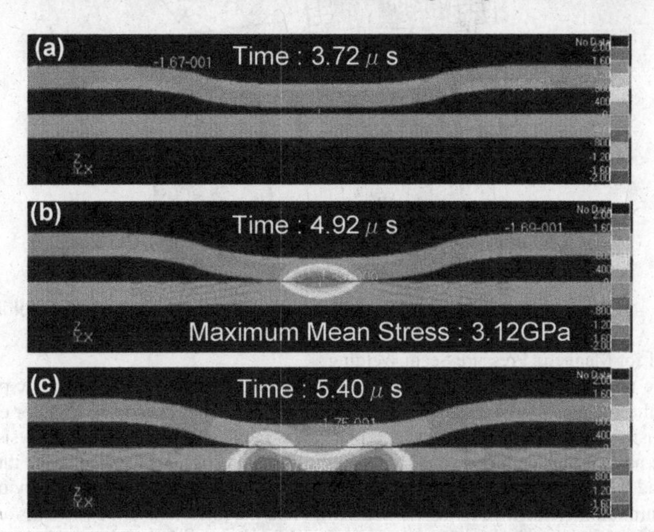

Fig. 3 Deformations and mean stress distributions of magnetic pressure seam welding. (collision velocity : 200 m/s)

Table 1 Material properties used for numerical analyses.

	Fe	Al
Young's Modulus (GPa)	206	70.3
Yield Stress (MPa)	500	200
Density (kg/m³)	7.87×10^3	2.70×10^3
Poisson's Ratio	0.3	0.345
Linear Expansion Coefficient (1/K)	1.18×10^{-5}	2.39×10^{-5}
Shear Modulus (GPa)	79.2	26.0
Specific Heat (J/kg·K)	440	900
Melting Point (K)	1808	933

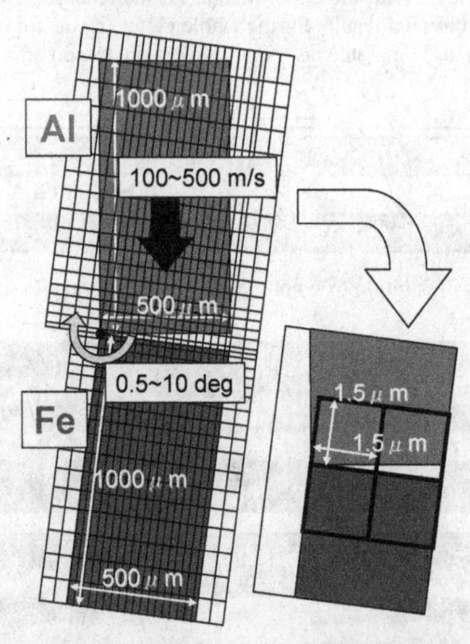

Fig. 4 Schematic illustration of partial model for magnetic pressure seam welding.

Partial Model for Magnetic Pressure Seam Welding

Since a wavy morphology was observed at the joint interface of the magnetic pressure seam welding and the cycle of wave is near 100 μm [7-10], more fine meshes have to be used for examining the joint mechanism precisely. So, a part of two plates was modeled for the precise analysis as shown in Fig. 4, where not only the collision speed but also the collision angle was varied in the range from 100 to 500 m/s and 0.5 to 10 degree, respectively according to the experimental and the previous numerical results. A minimum element size was $1.5 \times 1.5\ \mu m^2$, and the total number of elements and nodes was 281445 and 566956, respectively. The properties used for the partial model were the same as those in the previous computations for the whole components.

RESULTS AND DISCUSSIONS

Temperature Rise

One possible mechanism of the magnetic pressure seam welding seems to be an occurrence of local melting at the joint interface caused by large collision velocity. Since, in these numerical analyses, only the plastic strain would generate the temperature increment, the plastic strain occurred near the joint interface was examined. The maximum amount of plastic strain computed was the range from 0.27 and 1.65. A total energy per unit volume caused by the plastic strain can be written by the product of yield stress σ_Y and plastic strain ε^p in the case without work hardening. So, the temperature rise can be written by the following equation,

$$\frac{\sigma_Y \cdot \varepsilon^p}{c \cdot \rho} \tag{1}$$

Where c and ρ are the specific heat and the density. From the above equation, it is found that 1.0 plastic strain can generate a temperature rise of only 80.5 K for Al according to Table 1. So, a possible maximum temperature increment would be in the range from 22 to 132 K and any occurrence of local melting could not be considered because the melting temperature of Al is 933 K.

Pressure (Mean Stress)

From the previous analyses using the whole model, it was found that the very large mean stress occurred at collision point. This mean stress at the joint interface could be considered as a pressure at the joint surface. So, the influences of collision velocity and collision angle on the pressure were studied using the partial model. As same as the cases for the whole model, the pressure was locally applied at the joint interface and the point having the maximum pressure moved along the joint interface. Figure 5 shows the effects of collision velocity and collision angle on the maximum pressure. From this figure, it was found that the maximum pressure was 5 and 100 times larger than the yield stress of Al and the higher collision velocity could mostly generate the higher maximum pressure at the same collision angle. Also, it was revealed that the maximum pressure of each collision velocity would have a maximum value at a different collision angle.

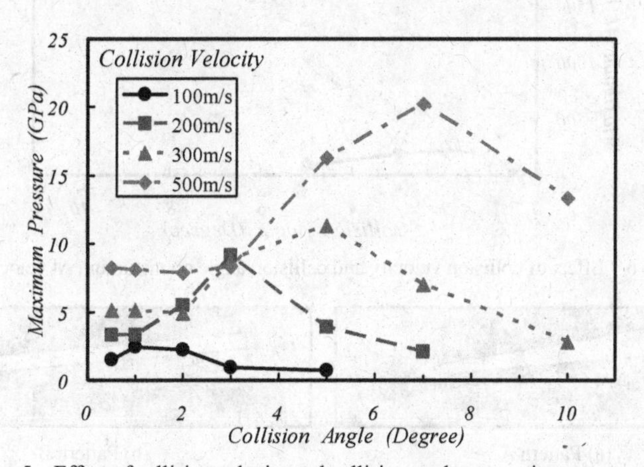

Fig. 5 Effect of collision velocity and collision angle on maximum pressure.

Al Velocity Parallel To Interface

Since the metal jet whose composition was mainly Al was observed experimentally, the Al velocity parallel to the joint interface was examined. The influences of collision velocity and collision angle on maximum Al velocity at joint interface were summarized into Fig. 6. The maximum Al velocity sometimes exceeded the collision velocity and such high Al velocities might be caused by the local high pressure at the joint interface. So, it can be considered that the differences between collision velocity and maximum Al velocity might generate the metal jet of Al. From Fig. 6, it was also found that the maximum Al velocity increased with increasing the collision angle, and achieved to the maximum value or almost saturated. Moreover, it was revealed that the maximum or saturated value of the maximum Al velocity at higher collision velocity occurred at large collision angle.

Plastic Strain Distribution

Although the plastic strain occurred near the joint interface would be much smaller for generating the local melting, two types of plastic strain distribution were obtained in these serial computations by varying the collision velocity and collision angle. Figs. 7(a) and (b) were typical examples of the plastic strain distributions generated near the joint interface and the influences of collision velocity and collision angle on the plastic strain distributions were summarized into Table 2. As shown in Fig. 7(a) which is denoted as pattern A, when the collision angle is smaller and the collision velocity is larger, the plastic strain near the joint interface decreased toward the end. On the other hand, in the other cases, large plastic strain continued over the whole joint interface as shown in Fig. 7(b) which is denoted as pattern B.

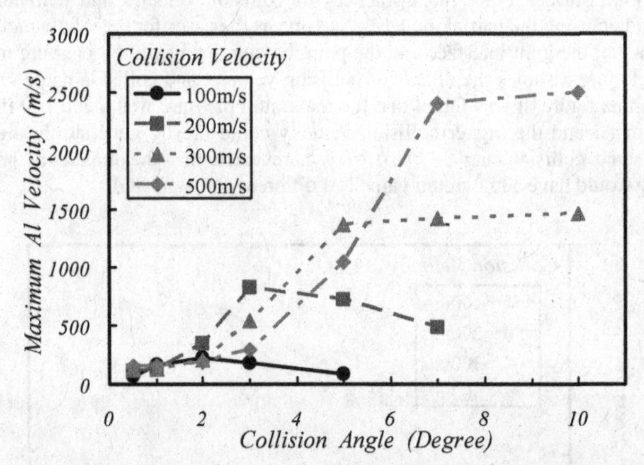

Fig. 6 Effect of collision velocity and collision angle on maximum Al velocity.

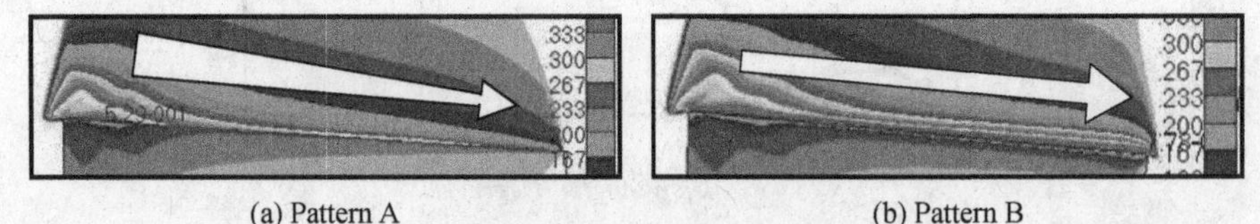

(a) Pattern A (b) Pattern B

Fig. 7 Plastic strain distributions near joint interface.

Table 2 Effect of collision velocity and collision angle on pattern of plastic strain distribution.

Collision Angle	Collision Velocity			
	100 m/s	200 m/s	300 m/s	500 m/s
0.5 degree	A	A	A	A
1 degree	A	A	A	A
2 degree	B	A	A	A
3 degree	B	B	A	A
5 degree	B	B	B	A
7 degree	-	B	B	B
10 degree	-	-	B	B

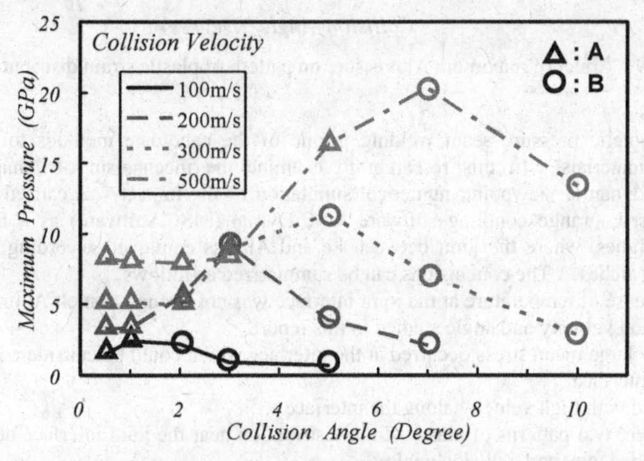

Fig. 8 Effect of maximum pressure on pattern of plastic strain distribution.

This difference in plastic strain distribution seems to be related to the effects of collision velocity and collision angle on the pressure and the Al velocity as shown in Figs. 8 and 9. From these figures, it was found that, before the maximum pressure and the maximum Al velocity achieved to the maximum value or almost saturated, the plastic strain distribution became to be the pattern A. So, it can be considered that, in these cases, the movement of Al along the interface might be prevented by the continuous contact between Al and Fe although a relatively large pressure was occurred. While, in the other cases (pattern B), Al could move along the joint interface before the growth of new contact and then the maximum Al velocity achieved to the maximum value or saturated. Since it was reported that the appropriate collision velocity and collision angle should be needed to create the joint interface in the magnetic pressure seam welding from the previous experimental studies [10,12,13], the plastic strain distribution near the joint interface might be related to the success of magnetic pressure seam welding. Also, from Figs. 8 and 9, it may be seen that the higher collision velocity would need the higher collision angle in order to develop the plastic strain distribution like pattern B.

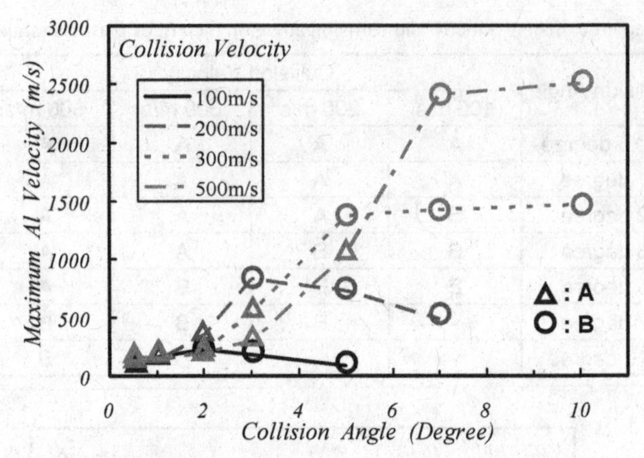

Fig. 9 Effect of maximum Al pressure on pattern of plastic strain distribution.

CONCLUSIONS

The magnetic pressure seam welding is one of the candidate methods to join thin sheet multifunctional materials. In this research, to examine the mechanism of magnetic pressure welding from a dynamic viewpoint, numerical simulation of the impact was carried out by using a commercial Euler-Lagrange coupling software MSC.Dytran (MSC.Software) as a first step of the computational studies, where the joint between Fe and Al was employed according to the previous experimental researches. The conclusions can be summarized as follows.

(1) The increase of temperature at the joint interface was not enough to melt Al or Fe in the range of collision velocity and angle studied in this report.

(2) The very large mean stress occurred at the interface which could be considered as the pressure at joint interface.

(3) Al moved with high velocity along the interface.

(4) There were two patterns of plastic strain distribution near the joint interface depending on the collision velocity and collision angle.

(5) The plastic strain pattern might be related to the success of magnetic pressure seam welding.

REFERENCES

[1] H. Serizawa, Y. Kawahito, H. Ogiwara, H. Tanigawa and S. Katayama, "Weldability of Reduced Activation Ferritic/Martensitic Steel under Ultra Power Density Fiber Laser Welding", *Proceedings of the 13th International Conference on Fusion Reactor Materials*, CD-ROM (2007).

[2] H. Tanigawa, T. Hirose, K. Shiba, R. Kasada, E. Wakai, H. Serizawa, Y. Kawahito, S. Jitsukawa, A. Kimura, Y. Kohno, A. Kohyama, S. Katayama, H. Mori, K. Nishimoto, R.L. Klueh, M.A. Sokolov, R.E. Stoller and S.J. Zinkle, "Technical Issues of Reduced Activation Ferritic/Martensitic Steels for Fabrication of ITER Test Blanket Modules", *Fusion Engineering and Design*, (2008), to be published.

[3] N. Iwamoto, M. Yoshida, S. Tanabe, T. Takeuchi and M. Makino, "Diffusion Welding of Mild Steel to Aluminum", *Transactions of JWRI*, **4**, 67-70 (1975).

[4] M. Kikuchi, H. Takeda and S. Morizumi, "Bonding Interfaces in Friction-and Explosive-Welded Aluminum and Steel Joints", *Journal of Japan Institute of Light Metals*, **34**, 165-172 (1984).

[5] T. Shinoda, M. Ogawa, S. Endo and K. Miyahara, "Friction Welding of Aluminum and Plain Low Carbon Steel", *Quarterly Journal of the Japan Welding Society*, **18**, 365-372 (2000).

[6] T. Aizawa, K. Okagawa and M. Kashani, "Seam Welding Method Using Magnetic Pressure from One Side", *Proceedings of International Symposium on Joining Technologies in Advanced Automobile Assembly 2005*, 97-105 (2005).
[7] M. Watanabe, S. Kumai and T. Aizawa, "Interfacial Microstructure of Magnetic Pressure Seam Welded Al-Fe, Al-Ni and Al-Cu Lap Joints", *Materials Science Forum*, **519-521**, 1145-1150 (2006).
[8] K.J. Lee, S. Kumai, T. Arai and T. Aizawa, "Interfacial Microstructure and Strength of Steel/Aluminum Alloy Lap Joint Fabricated By Magnetic Pressure Seam Welding", *Materials Science and Engineering A*, **471**, 95-101 (2007).
[9] S. Kumai, K.J. Lee and M. Watanabe, "Characteristic Interfacial Microstructure of Aluminum Alloy/Steel Lap Joints Fabricated by Several Advanced Welding Methods", *Proceedings of 11th International Conference on Aluminum Alloys*, **2**, 1945-1951 (2008).
[10] M. Watanabe, S. Kumai, K. Okagawa and T. Aizawa, "In-situ Observation of Magnetic Pulse Welding Process for Similar and Dissimilar Lap Joints Using a High-Speed Video Camera", *Proceedings of 11th International Conference on Aluminum Alloys*, **2**, 1992-1997 (2008).
[11] MSC.Software, *MSC.Dytran Manuals*, MSC.Software, (2008).
[12] M. Watanabe, S. Kumai, K. Okagawa and T. Aizawa, "In-situ Observation of the Magnetic Pulse Welding Process Using a High-Speed Video Camera", *Preprints of the National Meeting of Japan Welding Society*, **82**, 122-123 (2008).
[13] B. Crossland, *Explosive Welding of Metals and its Application*, Clarendon, Oxford (1982).

JOINING OF SILICON NITRIDE BY SLURRY OR PASTE

Naoki KONDO, Hideki HYUGA, Hideki KITA
National Institute of Advanced Industrial Science and Technology (AIST)
Shimo-shidami 2266-98, Moriyama-ku, Nagoya 463-8560, Japan

ABSTRACT

Porous or dense silicon nitrides were joined by silicon slurry, silicon nitride slurry or silicon paste, followed by sintering to obtain dense joined bodies. Combination of porous silicon nitride and silicon slurry exhibited average strength of more than 400 MPa. High powder packing density by slip cast like procedure and very low shrinkage by reaction bonding (RB) of the joint part resulted in the good strength.

INTRODUCTION

Joining technology is critically important for fabricating large scaled or complicated shaped ceramic components. This is also a key approach to combine and integrate several parts of different phases or different materials.

A lot of joining techniques have been developed to join silicon nitride (SN).[1] They were joined by using insert materials, such as silicon,[2] active metals or glasses. In the case of joining without insert materials, it was necessary to prepare smooth surface and apply mechanical pressure.

We have tried to develop a novel and unique joining technique without insert materials and mechanical pressure. Porous or dense silicon nitrides were joined by silicon slurry, silicon nitride slurry or silicon paste, followed by sintering to obtain dense joined bodies. By using this technique, silicon nitride original parts were joined by silicon nitride joint parts with the same composition to the original parts. Microstructures and strength of the joined bodies will be reported.

EXPERIMENTAL

Four types of combination were tried. Three types of silicon nitrides, i.e., porous RBSN, porous SN and dense SN, were prepared as original parts. Three types of slurry/paste, i.e., Si slurry, Si paste and SN slurry, were prepared to join original parts. These slurry/paste formed joint part between the original parts after sintering. The combinations are summarized in Table 1. Composition of all the original parts as well as joint parts were Si_3N_4 - 5 mass.% ZrO_2 - 5 mass.% $MgAl_2O_4$. They were after reaction bonding if the parts were fabricated from Si based powders.

Porous RBSN was fabricated from mixed powder of silicon and additives. The mixed powder was cold isostatically pressed (CIP) and reaction bonded at 1450°C for 2h in 0.1 MPa nitrogen atmosphere. Porous SN was fabricated from mixed powder of silicon nitride and additives. The mixed powder was CIPed and soaked at 1450°C for 2h in 0.1 MPa nitrogen atmosphere. Dense SN was fabricated from mixed powder of silicon nitride and additives. The mixed powder was CIPed and sintered at 1750°C for 6h in 0.5 MPa nitrogen atmosphere. Fabricated original parts were provided for joining. The planes for joining were grinded by #220 emery paper (for porous) or #200 whetstone (for dense).

Si slurry was fabricated by adding water and dispersant in Si mixed powder. Volume fraction of the mixed powder in the slurry was about 58 %. Si paste was fabricated by adding binder in the above Si slurry. By this procedure, the slurry increased viscosity and lost fluidity. SN slurry was fabricated by adding water and dispersant in SN mixed powder. Volume fraction of the mixed powder in the slurry was about 52 %.

In the case of joining by slurry, porous original parts were settled side by side with gap of about 2 mm, and the slurry was poured into the gap. The porous original parts soaked up water from the slurry like slip cast, thus consolidated joint part was formed in the gap. In the case of joining by paste, the planes with paste were joined.

Joined bodies with Si joint part were reaction bonded at 1450°C for 2h in 0.1 MPa nitrogen, followed by post sintering (PS) at 1800°C for 8h in 0.5 MPa nitrogen. Joined bodies with SN joint part were sintering at 1800°C for 8h in 0.5 MPa nitrogen. By these procedures, dense joined bodies were fabricated.

Microstructures of the specimens were observed by optical micrography and scanning electron micrography. Specimens (3 x 4 x 40 mm) for strength test were cut from the joined bodies. Stress direction was perpendicular to the joint part. Strength test was conducted under four point bending condition with outer and inner spans of 30 and 10 mm, and loading rate of 0.5 mm/min.

RESULTS AND DISCUSSION

Measured strengths of the joined bodies are summarized in Table 1. The combination, Porous RBSN - Si Slurry, showed the highest average strength of 404 MPa. On the other hand, the combination, Dense SN - Si Paste, showed only 172 MPa.

Table 1 Combination of original and joint parts, and average strength of joined body.

Original Parts	Joint Part	Major Fracture Origin	Average Strength
Porous RBSN	Si Slurry	Interface	404 ± 46
Porous SN	SN Slurry	Interface	238 ± 29
Porous RBSN	Si Paste	Joint part	263 ± 34
Dense SN	Si Paste	Interface	172 ± 20

Figure 1 shows the joined specimen of Dense SN - Si Paste. Thickness of the joint part was about 20 μm. Large pores were found at the joint part as indicated by arrows. These pores were also observed on the fractured surface. As the dense silicon nitride did not soak up water from Si paste, paste seemed to be separated into Si joint part and standing water during drying. Formation of standing water resulted in pore. Fracture occurred at the interface between the original and formed joint parts. As the original parts were already densified, diffusion at the interface did not seem to be so active. These two are the reasons why this combination did not show good strength.

In the case of the combination, Porous RBSN - Si Paste, no significant pores were found at the joint part by visual inspection. But pores still exist as blots were found by ink check. Fractured surface of this combination is shown in Fig. 2. Large flaws like Fig. 1 were not observed, however, densification of the joint part was insufficient. As original parts were porous and can soak up water from the paste, water separation was avoided and powder packing in the consolidated joint part became close. This behavior is similar to slip cast. However, low fluidity of the paste prevented achieving high powder packing. Therefore pores remained in the joint part. Only in this combination, fracture mainly occurred at the joint part. As this combination used porous RBSN and silicon powder, mass transfer was enhanced during RB and PS. Additionally, very low shrinkage (<1%) after RB is advantageous to reduce shrinkage mismatch between original and joint parts. These resulted in strong interface connection between original and joint parts, and the fracture seemed to occur from the residual pores in the joint part.

In the case of the combination, Porous SN - SN Slurry, thickness of the joint part was about

1.5 mm. This joint part was formed by consolidation of slurry like slip cast. Some cracks perpendicular to the joint interface were found by visual inspection. These cracks seemed to form during drying of the consolidated joint part. The cracks did not affect the strength, since loading direction was parallel to the crack direction. Blots by ink check were found, which mean existence of pores. These pores were observed in fractured surface as shown in Fig. 3. The slurry used for this combination has a potential to make fully densified body under the same PS condition. One possible reason of insufficient densification is that the consolidation condition differed from usual slip casting. Fracture mainly occurred at the interface between original and joint parts. As this combination used SN slurry, the advantages of RB were not expected. Thus, interface connection was not so strong, resulted in the fracture at interface.

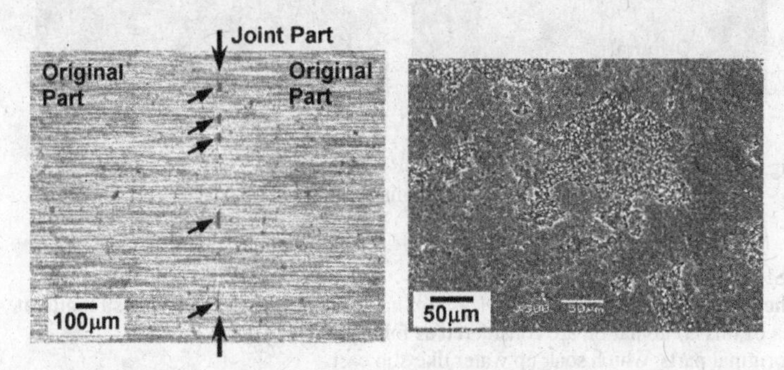

Fig. 1 The combination of Dense SN - Si Paste.
View of the joined specimen (Right) and fractured surface (Left)

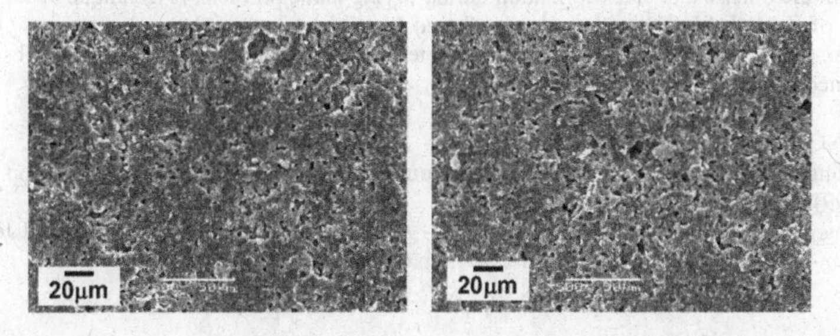

Fig. 2 (Left) The combination of Porous RBSN – Si Paste. Fractured surface.
Fig. 3 (Right) The combination of Porous SN – SN Slurry. Fractured surface.

Joined specimen, Porous RBSN - Si Slurry, is shown in Fig. 4. In this combination also, some of the specimen had cracks perpendicular to the joint interface. Therefore, a technique to suppress the formation of crack during drying is needed when slurry is used for joining. Except for the cracks, no other flaws were detected by ink check. Observation even in higher magnification did not find flaws at the interface as well as joint part. Therefore, joint part with high powder packing seemed to form by consolidation of slurry. This combination showed the highest average strength of 404MPa among the four combinations. However, fracture was mainly occurred at the interface. This indicates residual flaws still exist at the interface.

Fig. 4 The combination of Porous RBSN - Si Slurry.
View of the joined specimen (Right) and joined interface (Left)

CONCLUSIONS

The combination, Porous RBSN - Si Slurry, achieved the average strength of 404MPa. Advantages of this combination are considered as follows.
1) Porous original parts, which soak up water like slip cast.
2) Si slurry, which has good fluidity and forms joint part with high powder packing density.
3) Si joint part, which enhances mass transfer during RB
4) Reaction bonding, which accompanies very low shrinkage (<1%)

On the other hand, crack formation during drying is the problem. A technique to suppress the formation of crack during drying must be developed.

As this combination seemed to have a potential to achieve strong joint, further observations of this joined body are under progress.

REFERENCES

[1] Loehman RE, "Recent Progress in Ceramic Joining," Key Eng. Mater., Vol.161-163, pp.657-661, (1999).
[2] Eiling A and Passing G, "Si - Nanopowder - A Material for Synthesis of Si_3N_4 and Joining of Si_3N_4 - Parts," Key Eng. Mater., Vol.89-91, pp.67-71, (1994).

SEGREGATION MECHANISM IN (M=Al, Ga) $Zn_{1-x}M_xO$ CERAMICS AND ITS INFLUENCE ON THE THERMOELECTRIC PROPERTIES

J. P. Wiff, Y. Kinemuchi and K. Watari
National Institute of Advanced Industrial Science and Technology (AIST)
463-8560 Nagoya, Aichi, Japan

ABSTRACT

This work focuses on the segregation and Hall mobility of Al-doped ZnO (Al-ZnO) and Ga-doped ZnO (Ga-ZnO) ceramics. It was found that their Hall mobilities are quite different in spite of their similar thermoelectric properties. Potential barriers or carrier traps at the grain boundaries can consistently explain the Hall mobility behavior as a function of carrier concentration in Al-ZnO samples. On the contrary, the influence of potential barrier on Hall mobility was not dominant in Ga-ZnO samples. TEM and EDX analyses of Ga-ZnO indicate no preferential segregation at the grain boundary, which is responsible for the less influence of potential barrier on carrier transport in Ga-ZnO ceramics. The results indicate that the high Hall mobility detected at low carrier concentration is useful for decreasing the electric resistivity and increasing the Seebeck coefficient, thus enhancing the power factor of doped ZnO ceramics.

INTRODUCTION

ZnO has been studied as thermoelectric material due to its wide band gap, good chemical stability at high temperatures and environmental friendliness[1,2]. However, its high thermal conductivity reduces the efficiency in terms of figure of merit *ZT*, $ZT = \frac{\alpha^2}{\rho\kappa}T$, where α, ρ, κ, and *T* are Seebeck coefficient, electrical resistivity, thermal conductivity, and absolute temperature, respectively. Several dopants have been added to ZnO in order to decrease the electrical resistivity[1,3,4]. However, depending of the solubility limit, at high concentration of dopant a secondary phase like spinel is formed[5,6].

Al doping has been used for improving the electric conductivity in ZnO ceramics[1,7,8]. At Al addition lower than 1 *at%* the low Hall mobility (μ) deteriorates the thermoelectric properties. Normally the best thermoelectric performance is obtained at Al additions between 1 and 3 *at%*[7,8,9]; however under those concentrations a secondary spinel phase ($ZnAl_2O_4$) is formed. In general, Hall mobility of Al-added ZnO (Al-ZnO) tends to increases as the carrier concentration increases. This behavior can be explained considering the influence of potential barriers at the Al-ZnO grain boundaries on the Hall mobility[10]. Recently, Al segregation was found at the grain boundaries causing a Zn deficiency that forms potential barriers[11].

Ga and Al are in the same group in the periodic table and their ionic radii are similar, so it is expected that both Al-ZnO and Ga-ZnO ceramics show similar thermoelectric properties. Ga-ZnO has been widely studied as thin film for use in optical applications; however, there are few studies about its use as a thermoelectric material[12,13,14], especially as bulk, because at high Ga additive concentrations its density is drastically reduced[12,14].

An interesting behavior in Ga-ZnO ceramics is their high Hall mobility at low carrier concentration (n) in opposition to that observed in Al-ZnO ones[9]. It is expected that at high Hall mobility and reducing the carrier concentration the Seebeck coefficient, the electrical resistivity and subsequently the thermoelectric properties of ZnO ceramics can be enhanced. Therefore, this work studies the relation between ion segregation and Hall mobilities of Al-ZnO and Ga-ZnO and its influence on the thermoelectric properties.

EXPERIMENTAL

$Zn_{1-x}M_xO$ samples with M={Al or Ga} and x={0.002, 0.005 or 0.01} were prepared by solid-state reaction of ZnO (99.8%, Hakusui Tech. Co. Japan), γ-Al_2O_3 (AKP-G015, 99.995%, Sumimoto Chemical, Japan) and Ga_2O_3 powders (99.9%, Kanto Reagents, Japan).

Fifty grams of ZnO and M-type oxide powders were mixed in stoichiometric quantities in a polyethylene bottle containing 80 ml of ethanol and 130 g of ZrO_2 balls at 100 rpm for 20 h. Afterwards, ethanol and ZrO_2 balls were removed in a rotary evaporator and the slurry was dried under vacuum (~1 Pa) at 60°C for 24 h. Dried powders were sieved using a mesh #250.

Samples were compacted in pellets of 5 mm of thickness and 30 mm in diameter. All samples were uniaxially pressed at 30 MPa for 30 s and then hydrostatically cold pressed at 100 MPa for 1 min. Finally, all samples were sintered in air at 1400°C for 10 h at a heating rate of 10°C/min.

All samples were analyzed using an X-ray diffractometer (Rigaku RINT2000, Japan) with CuK_α radiation at room temperature. Before thermoelectrical characterization, all samples were cut in pieces of 5 mm × 5 mm × 14 mm approximately. Afterward, all faces were polished to ensure the reproducibility of measurements. Seebeck coefficient and electrical resistivity were measured by static DC method (ULVAC-ZEM-1, Japan) under He atmosphere from room temperature up to 800°C and temperature gradients of 20°C, 30°C, and 40°C. All samples were measured twice and the reproducibility of Seebeck coefficient and electrical resistivity were confirmed. Carrier concentration and Hall mobility were analyzed by DC Hall measurement at room temperature (Resistest 8300, Toyo-Technica, Japan) under a magnetic field of 0.75 T. Samples had a square shape of 10 mm × 10 mm × 0.10 mm in size. Four platinum electrodes were sputtered at the corners of the square for the Van der Pauw method. In all cases the ohmic behavior at the electrode was verified and the Hall voltage was at least two orders of magnitude stronger than noise voltage.

RESULTS AND DISCUSSION

Figures 1a and 1b correspond to X-ray diffraction patterns at room temperature of $Zn_{1-x}Al_xO$ and $Zn_{1-x}Ga_xO$ x={0.002, 0.005, or 0.01} samples, respectively. Both, Al-ZnO and Ga-ZnO samples contain a secondary phase: $ZnAl_2O_4$ (JCPDF #05-0669) and $Zn_9Ga_2O_{12}$ (JCPDF #50-0448), respectively.

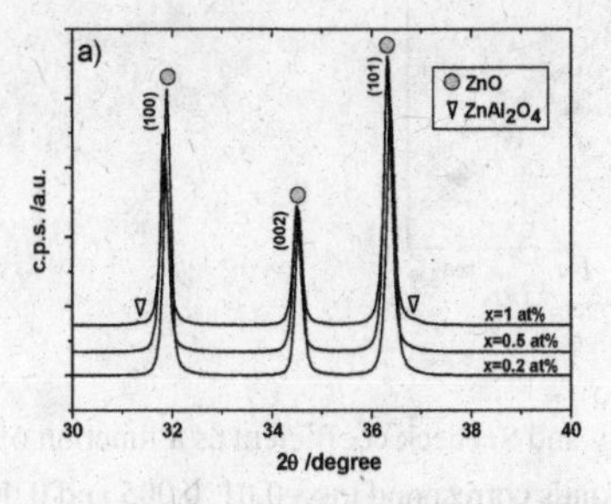

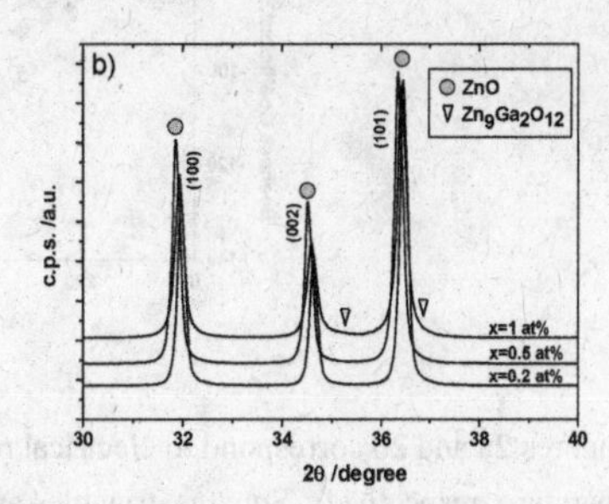

Figure 1 XRD of a) Al-ZnO and b) Ga-ZnO samples at room temperature

Al-ZnO samples exhibited densities over 99% whereas in Ga-ZnO the density slightly decreased as the Ga addition increased. In Ga-ZnO samples with x=0.01 (1 *at*%) the minimum density of 96% was obtained. In spite of the low density of Ga-ZnO samples their electric resistance was permanently low compared with Al-ZnO (Figure 2a) while their Seebeck coefficients follow similar tendency. However, in Ga-ZnO samples with low additive concentration an enhancement of Seebeck coefficient was observed (Figure 2b, open diamond).

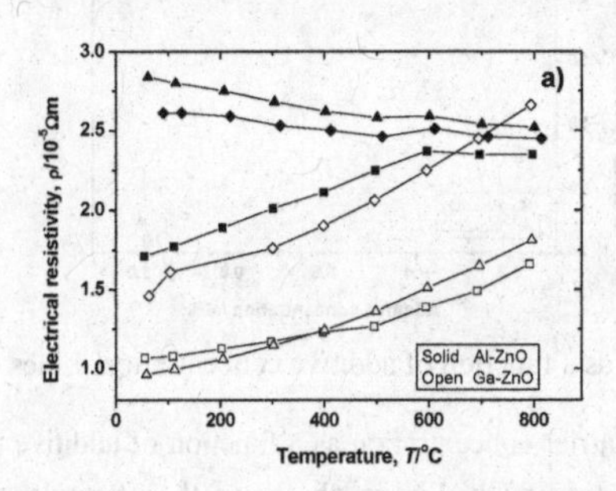

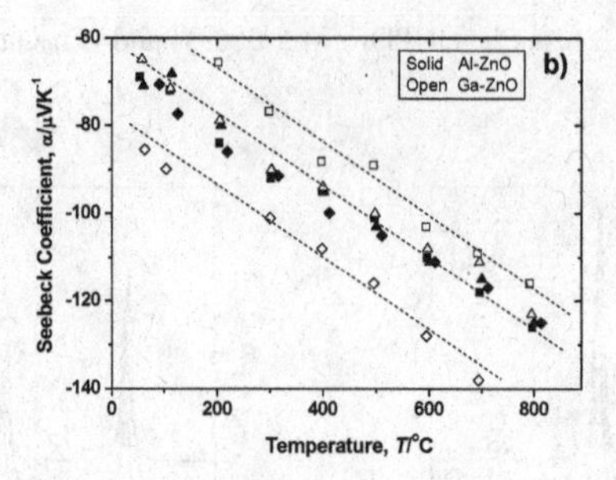

Figures 2a and 2b correspond to electrical resistivity and Seebeck coefficient as a function of temperature, respectively. Squares, triangles and diamonds correspond to x=0.01, 0.005 and 0.002, respectively. Lines are used only as a reference

Figure 3 shows the thermal conductivity of Al and Ga-ZnO samples at room temperature. Regarless of Al addition, Al-ZnO showed similar thermal conductivity; whereas in Ga-ZnO the thermal conductivity gradually decreases as the Ga addition increased.

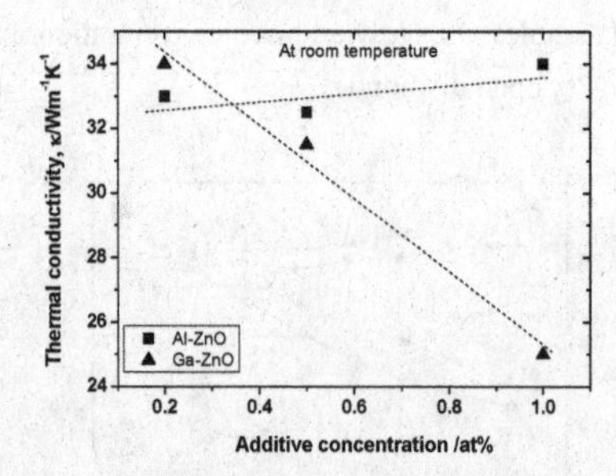

Figure 3 Thermal conductivity as a function of additive concentration. Lines are used only as a reference

Figure 4a shows the carrier concentration as a function of additive concentration of Al or Ga, indicating a high activation ratio of Ga. Figure 4b shows the effective mass (m^*) of Al-ZnO and Ga-ZnO samples as a function of carrier concentration at room temperature. The procedure of estimating the effective mass has been described elsewhere[8]. The effective mass slightly increases with the carrier concentration in agree with previous observations[8]. Figure 4c shows the Hall mobility as a function of carrier concentration at room temperature. Al-ZnO and Ga-ZnO samples exhibit opposite behaviors: in Al-ZnO samples Hall mobility increases with the carrier concentration, whereas in Ga-ZnO samples Hall mobility decreases as the carrier concentration increases.

At low carrier concentration the Hall mobility in Ga-ZnO samples is higher with respect to that in Al-ZnO samples (Figure 4c). This fact induces a significant reduction of electrical resistivity and an increment of Seebeck coefficient (Figure 2), thus enhancing the power factor of Ga-ZnO with respect to the Al-ZnO samples.

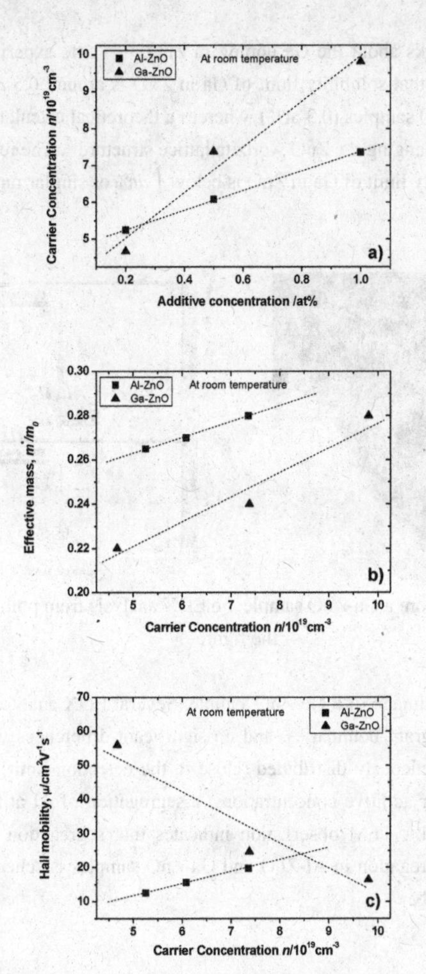

Figure 4a Carrier concentration as a function of additive concentration. Figures 4b and 4c correspond to effective mass and Hall mobility as a function of carrier concentration. Lines are used only as a reference.

The Hall mobility tendency of Al-ZnO ceramics can be explained by the carrier traps being preferentially located at grain boundaries generating a potential barrier in the path of carriers. At low carrier concentration the small proportion of carriers trapped at the grain boundaries is compensated respect to that at high carrier concentration. Thus, at low carrier concentration the Hall mobility is low and vice versa.

There are some works about the Ga doping in ZnO[5,15]. Some experimental works based on XRD measurements suggest that solubility limit of Ga in ZnO is around 0.5 *at%*[5], quite similar with the reported value for Al-ZnO samples (0.3 at%), whereas a theoretical calculation predicts a solubility limit up to 12 *at%* without changing the ZnO wurtzite lattice structure[15]. The results showed in Figures 1b and 2 indicate the solubility limit of Ga in ZnO is below 1 *at%* or similar range as reported by Yoon *et al.*

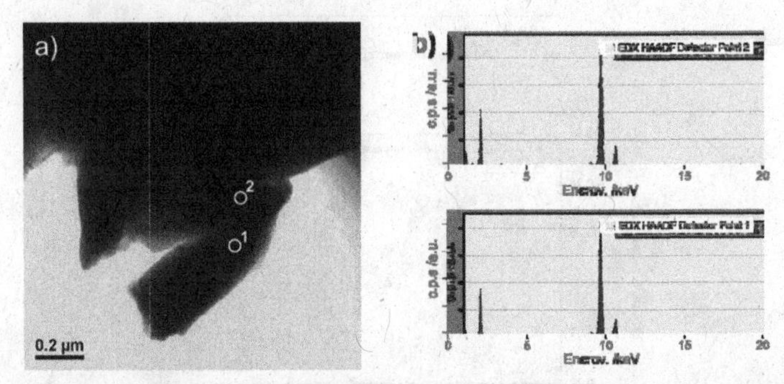

Figure 5a HRTEM image from a Ga-ZnO sample. 5b EDX analysis from points 1 and 2 indicated in the figure

Figure 5a is a TEM image of a Ga-ZnO sample. Several EDX analyses were performed both in the (1) grains and at (2) grain boundaries, and no significant differences were found (Figure 5b), indicating that Ga is homogeneously distributed (close to the detection limit). On the other hand, in Al-ZnO samples with similar additive concentrations, a segregation of Al at the grain boundaries as thin layer has been reported[11]. TEM observation indicates that segregation of Al and Ga in ZnO samples is different. The segregation in Al-ZnO and Ga-ZnO samples is schematically represented in Figures 6a and 6b, respectively.

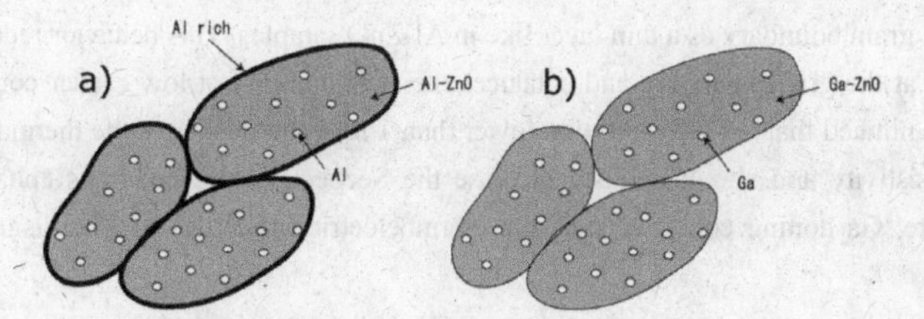

Figure 6 Schematic representation of segregation in a) Al-ZnO and b) Ga-ZnO samples

Figure 6a represents that in Al-ZnO samples the Al-rich thin layer around the Al-ZnO acts as a potential barrier for reducing the Hall mobility; whereas in Ga-ZnO samples, Ga is well dispersed and it does not segregate at the grain boundaries (Figure 6b), thus reducing the carrier traps at the grain boundaries and subsequently enhancing the Hall mobility especially at low carrier concentrations (Figure 4c).

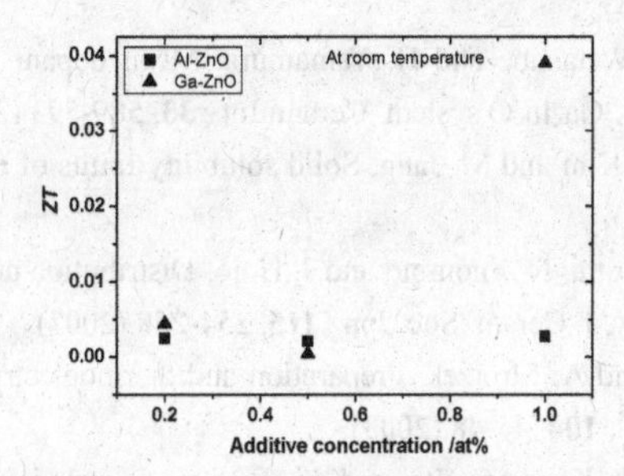

Figure 7 ZT index as a function of additive concentration

Figure 7 shows the ZT index at room temperature for Al-ZnO and Ga-ZnO samples. It is noticed that Ga-ZnO samples shows higher ZT index with respect to Al-ZnO samples due to the enhancement in thermal conductivity, electrical resistivity and Seebeck coefficient, which is caused both by Ga doping and by the Hall mobility-carrier concentration behavior. Therefore, the Ga looks like a promising dopant for improving the thermoelectric performance in ZnO-based thermoelectric materials with a minimum additive concentration.

CONCLUSIONS

The difference in Al and Ga-ZnO Hall mobility was discussed considering the segregation of dopant. TEM images suggest that Ga is well distributed into the ZnO matrix and it does not tend to segregate at the grain boundary as a thin layer like in Al-ZnO samples. This behavior reduces the effect of carrier traps at the grain boundary and enhances the Hall mobility at low carrier concentration. In addition, it was noticed that Ga concentration lower than 1 *at%* can decreases the thermal conductivity and electric resistivity and simultaneously increase the Seebeck coefficient, thus enhancing the ZT index. Therefore, Ga doping could enhance the thermoelectric properties of ZnO using a minimum concentration.

REFERENCES

[1]M. Ohtaki, T. Tsubota, K. Eguchi and H. Arai, High-temperature thermoelectric properties of $(Zn_{1-x}Al_x)O$, J. Appl. Phys., **79**, 1816-1818 (1996).

[2]H. Kaga, Y. Kinemuchi, H. Yilmaz, K. Watari, H. Nakano, H. Nakano, S. Tanaka, A. Makiya, Zenji Kato and K. Uematsu, Orientation dependence of transport property and microstructural characterization of Al-doped ZnO ceramics, Acta Mater., **55**, 4753-4557 (2007).

[3]K. Kim, S. Shim, K. Shim, K. Niihara and J. Hojo, Microstructural and thermoelectric characteristics of zinc oxide-based thermoelectric materials fabricated using a spark plasma sintering process, J. Am. Ceram. Soc., **88**, 628-632 (2005).

[4]K. Kakinuma, T. Shibo, M. Watanabe and H. Yamamura, Mean dopant ion radius dependency of electrical resistivity in the $Zn_{1-x-y}Ga_xIn_yO$ system. Ceram. Int., **33**, 589-593 (2007).

[5]M. Yoon, S. Lee, H. Park, H. Kim and M. Jang, Solid solubility limits of Ga and Al in ZnO, J. Mat. Sci. Lett., **21**, 1703-1704 (2002).

[6]K. Shirouzu, T. Ohkusa, M. Hotta, N. Enomoto and J. Hojo, Distribution and solubility limit of Al in Al_2O_3-doped ZnO sintered body, J. Ceram. Soc. Jpn., **115**, 254-258 (2007).

[7]K. Cai, E. Muller, C. Drašr and A. Mrotzek, Preparation and thermoelectric properties of Al-doped ZnO ceramics, Mat. Sci. Eng. B, **104**, 45-48 (2003).

[8]J. P. Wiff, Y. Kinemuchi, H. Kaga, C. Ito and K. Watari, Correlations between thermoelectric properties and effective mass caused by lattice distortion in Al-doped ZnO ceramics, Doi:10.1016/j.jeurceramsoc.2008.09.014.

[9]T. Tsubota, M. Ohtaki, K. Eguchi and H. Arai, Thermoelectric properties of Al-doped ZnO as a promising oxide material for high-temperature thermoelectric conversion, J. Mater. Chem., **7**, 85-90 (1997).

[10]V. Srikant, V. Sergo and D. Clarke, Epitaxial aluminium-doped zinc oxide thin films on sapphire: II, Effect of substrate orientation, J. Am. Ceram. Soc., **78**, 1935-1939 (1995).

[11]Y. Kinemuchi, H. Kaga, S. Tanaka, K. Uematsu, H. Nakano and K. Watari, Zinc oxide ceramics with high mobility as n-type thermoelectric materials, Mater. Sci. Forum, **561-565**, 581-586 (2009).

[12]B. Cook, J. Harringa and C. Vining, Electrical properties of Ga and ZnS doped ZnO prepared by mechanical alloying, J. Appl. Phys., **83**, 5858-5861 (1998).

[13]G. Paul and S. Sen, Sol-gel preparation, characterization and studies on electrical and

thermoelectrical properties of gallium doped zinc oxide films, Mater. Lett., **57**, 742-746 (2002).
[14]R. Wang, A. Sleight and D. Cleary, High conductivity in gallium-doped zinc oxide powders, Chem. Mater., **8**, 433-439 (1996).
[15]C. Ren, S. Chiou and C. Hsue, Ga-doping effects on electronic and structural properties of wurtzite ZnO, Phys. B, **349**, 136-142 (2004).

PRODUCTION OF NOVEL ARCHITECTURES THROUGH CONTROLLED DEGRADATION OF ELECTROSPUN PRECURSORS

Satya Shivkumar, Xiaoshu Dai
Department of Mechanical Engineering, Worcester Polytechnic institute
Worcester, MA 01609, USA

ABSTRACT

Fibers and highly porous architectures of hydroxyapatite were produced by controlled degradation of a scaffold of an electrospun precursor. The precursor was prepared by aging triethyl phosphite and calcium nitrate and was directly added to an aqueous solution of Polyvinyl alcohol (PVA). This electrospun structure was calcined at various temperatures to obtain a residual inorganic network, which consisted predominantly of hydroxyapatite. The effects of process temperature and time are investigated. These structures can have many potential uses in the repair and treatment of bone defects, drug delivery and tissue engineering.

INTRODUCITON

Sintered calcium phosphate ceramics based on hydroxyapatite (HA, $Ca_{10}(PO_4)_6(OH)_2$) or β-tricalcium phosphate (β-TCP, $Ca_3(PO_4)_2$) are one of the most prominent materials used in the treatment of orthopedic defects. They are also being used by many researchers for bone tissue engineering. Their chemical composition is highly related to the mineral phase of natural bone [1-3]. These similarities lead to a good biocompatibility between these materials and bone. Synthetic hydroxyapatite (HA) particles, films, coatings, fibers and porous skeletons are used extensively in several biomedical applications [4,5].

Traditionally, this bioceramic can be synthesized by solid state reactions, plasma techniques [6], hydrothermal hotpressing [7], and many wet chemical precipitation and mechano-chemical methods [8,9]. The sol-gel approach provides significantly easier conditions for the synthesis of HA and has recently attracted much attention. Sol-gel process refers to a low-temperature method using chemical precursors that can produce ceramics and glasses with better purity and homogeneity [10]. Compare to the conventional methods, the most attractive features and advantages of sol-gel process include (a) molecular-level homogeneity can be easily achieved through the mixing of two liquids; (b) the homogeneous mixture containing all the components in the correct stoichiometry ensures a much higher purity; and, (c) much lower heat treatment temperature to form glass or polycrystalline ceramics is usually needed without resorting to a high temperature. This process is becoming a common technique to produce ultra fine and pure ceramic powders, fibers, coatings, thin films, and porous membranes. More recently, the sol-gel method has been extensively developed and used in biotechnology applications [11].

Ceramic nanofibers synthesized via electrospinning were reported first in 2002 [12]; since then, various ceramic metal oxide fibers (with diameters between 200 and 400 nm) have been obtained by high temperature calcination of the precursor organic–inorganic composite nanofibers assembled by electrospinning [13]. In a number of electrospun ceramic systems, a relatively long aging time is required to achieve the complete mixing between the polymer and the ceramic content [14]. It was generally observed that the calcination temperature has a great influence on both the crystalline phase and the surface morphology of the fibers [15]. Many groups have reported that the morphologies of the final sample depend strongly on the calcination temperature [16]. Recently, Wu et al. [17] have produced HA fibers 10 to 30 µm in diameter by combining sol-gel processing with electrospinning. They used a precursor mixture of triethyl phosphite, calcium nitrite tetrahydrate and a polymer additive. After electrospinning and calcination at 600°C for 1 hr, HA fibers were obtained. The fiber diameters were in the order of 10~30 μm with a grain size of 1 μm. Further developments in this area aimed at reducing the fiber diameter in order to enhance the resorption rate in the body.

In this contribution, a hydroxyapatite fibrous network is produced by calcination of an electrospun PVA – calcium phosphate sol mixture. The effects of aging and calcinations conditions on the final phase of the ceramic product are studied.

EXPERIMENTAL

Triethyl phosphite (TEP, Aldrich, USA) and calcium nitrate tetrahydrate ($Ca(NO_3)_2{\cdot}4H_2O$, Aldrich, USA) were used as the raw materials for preparing the inorganic sol. PVA with various weight average molecular weights (M_W) was obtained from Aldrich Chemical Company, Milwaukee, WI. Distilled water was used as the solvent.

The inorganic sol was prepared by sol-gel routine described previously [18]. Appropriate amounts (11.9 g) of calcium nitrate tetrahydrate (to obtain a Ca/P ratio of 1.67) were dissolved in 10 mL of distilled water. About 5.2 mL of triethyl phosphite was hydrolyzed with 10 mL of distilled water. The nitrate solution was added dropwise into the hydrolyzed phosphate solution. The resulting mixture was stirred vigorously for 2 hr in an aqueous environment and aged at different temperatures for different time as shown in Table 1 (a). The sol used for preparing the electrospun precursor was aged at 80°C for 24 hr. A transparent sol was obtained after the aging treatment.

The sol was dried at 80°C for 24 hr to drive off all the solvent. A white gel was obtained after drying. The gel was calcined at temperatures between 300 - 800°C for various times. A white powder was generally obtained after calcination. The times and temperatures used are summarized in Table 1(b).

Table 1 Experimental Conditions used during the production of the sol (a) and the inorganic powder (b).

Aging Temperature, T_1(°C)	70	80
Aging time, t_1(hr)	24	
Calcination Temperature, T_2(°C)	700	
Calcination time, t_2(hr)	2	

(a)

Aging Temperature, T_1(°C)	70									
Aging time, t_1(hr)	36									
Calcination Temperature, T_2(°C)	300	400	500	600	700					800
Calcination time, t_2(hr)	0.5	0.5	0.5	0.5	0.5	2	3	4	5	0.5

(b)

PVA with a molecular weight (M_W) of 67,500 g/mol was purchased from sigma-aldrich, USA. Appropriate amounts of PVA were dissolved in distilled water to achieve a concentration of 15 wt.% . The dissolution was conducted in a water bath at 80°C, with constant stirring for at least 2 hr to ensure complete dissolution.

The electrospinning precursor was prepared by mixing the sol and the polymer solutions at room temperature. The volumetric ratio between the PVA solution and the inorganic sol was varied between 1:5 and 5:1 corresponding to sol volume fraction between 83.3vol% and 16.7vol%. The mixture was then sealed and vigorous stirring was continued for at least 8 hr at 25°C. After this, the stirring was stopped to facilitate the removal of gas bubbles.

The electrospinning apparatus consisted of a 1 mL syringe, equipped with an 18-gauge (Outer diameter = 1.27 mm, 52 mm long) stainless steel needle that were mounted horizontally in a syringe pump as shown in Figure 1. A potential of 20 kV was applied to the needle immediately after a pendant drop formed at the tip of the needle. The collector plate was covered with aluminum foil, weighed in advance, and positioned at a distance of 10 cm from the needle.

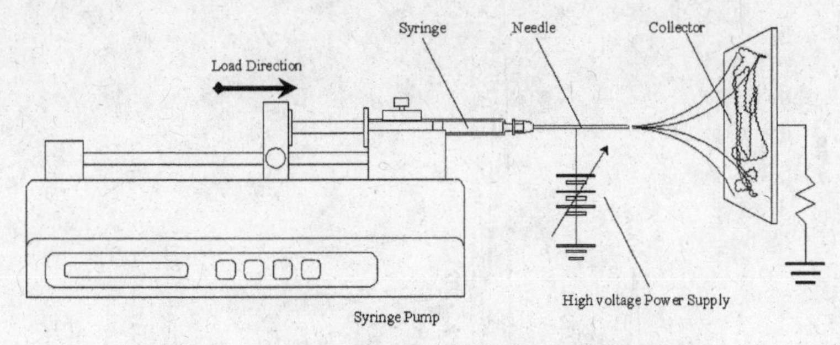

Figure 1. Schematic illustration of the experimental set-up.

The PVA/Sol mixtures obtained after electrospinning were dried at room temperature for at least 8 hr. The dried electrospun hybrid samples were calcined at 600°C for 6 hr.

The white powder obtained after calcination was analyzed by X-Ray diffraction (XRD) (Rigaku) with Cu K-α radiation. Also XRD analysis was conducted on small samples scraped off from the calcined electrospun specimens. The diffraction data were analyzed with JADE software. The electrospun samples were sputter coated with gold-palladium and examined in JSM-840 scanning electron microscope (SEM).

RESULTS AND DISCUSSION

Preparation of the Sol

A colorless sol was obtained after the aging process. When this sol was calcined, the structure consisted predominantly of HA and minor phases, such as tricalcium phosphate and CaO, in small fractions [12]. It has been reported that the aging time and temperature are critical in the formation of HA [33]. The effect of aging temperature and aging time are shown in Figure 2 and Figure 3, respectively.

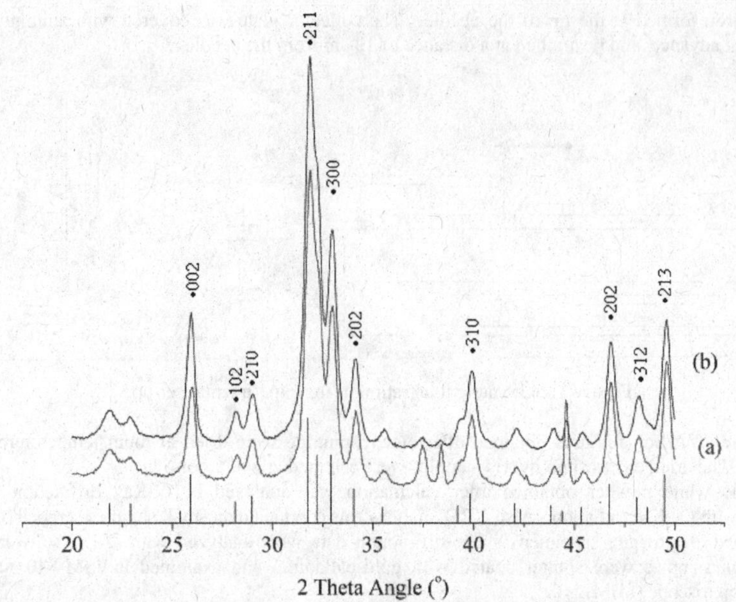

Figure 2. X-ray diffraction patterns for sols aged at (a) 70°C and (b) 80°C. After aging for 24 hr at the preceding temperatures, the sol was calcined at 700°C for 2 hr. The primary diffraction peaks for hydroxyapatite (•) are indicated.

It can be observed that the intensity of the characteristic peaks corresponding to HA phase in samples aged at 80°C is more pronounced than the samples aged at 70°C. It has been reported by several investigators researchers that higher aging temperatures favor of the formation of HA [19]. The aging time did not have a significant effect on the formation of HA (Figure. 3). In this case, there are no obvious distinctions between the characteristic peaks of HA phase within the two samples. The characteristic peaks at 2θ=29.8-31.1° indicate the presence of β- tricalcium phosphate (β-TCP) in the samples aged for 36 hr. This phenomenon is consistent with the data of Chai et al. [20], who reported that there is a trend towards the formation of β-TCP when a longer aging time is used. Thus, any aging time longer than 36 hr may not be desirable for the formation of HA.

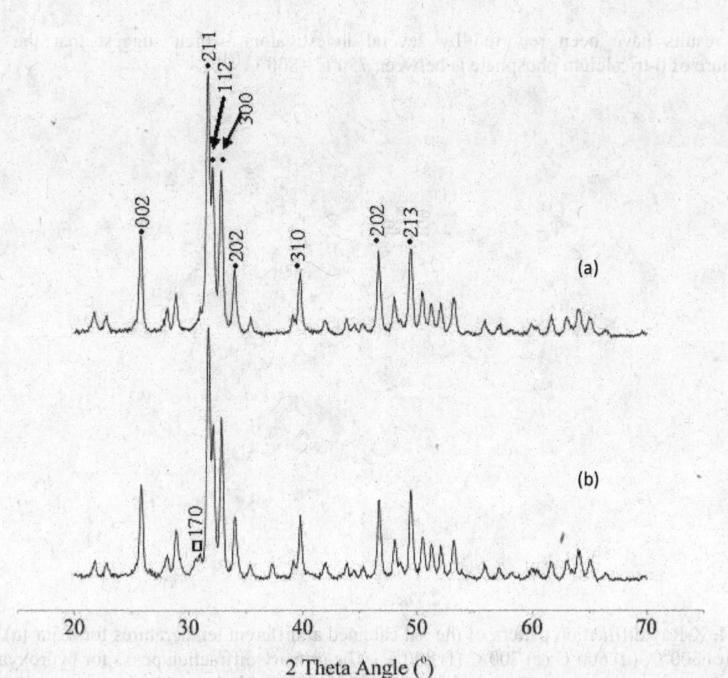

Figure 3. XRD pattern of 70°C aged sol for (a) 24 hr and (b) 36 hr. The samples are calcined at 700°C for 2 hr. The primary diffraction peaks for hydroxyapatite (•) and β-tricalcium phosphate (□) are indicated.

Liu et al. [12] reported that a minimum 8 hr aging is necessary to obtain HA and prevent the formation of the secondary phase, CaO. They also concluded that the fraction of HA increases and impurity phases, such as CaO, $Ca_2P_2O_7$, $Ca_3(PO_4)_2$, gradually disappear with increasing aging time. In the XRD patterns of both samples (Figure 3), there is no evident peak corresponding to the CaO phase. Thus, the formation of the CaO phase can be minimized by the aging process used in this investigation.

Calcination of the Gel

The sol was heat treated at 80°C to drive off the solvent and form a white gel. During this procedure, a weight loss about 50wt% was observed. This weight loss was mainly from the loss of the absorbed solvent (water and alcohol). The dried gel was calcined at various temperatures for various times. Upon calcination, an additional 40wt% weight loss was observed. This is primarily due to the decomposition of the nitrate salt and the loss of the combined water. The XRD patterns of the gel prepared from treating the sol at 70°C for 36 hr followed by calcination for 0.5 hr at various temperatures are shown in Figure 4. It can be seen that a highly crystalline form of HA is observed upon calcination. The characteristic peaks of HA phase were first observed in the sample calcined at 500°C (Figure 4(c)). This result is consistent with the data of Chai et al. [21] and Lopatin et al. [22], who reported that the temperature to form crystallized hydroxyapatite is between 420°C and 460°C. Some of the characteristic peaks in Figure 4 also correspond to the secondary phase – β-tricalcium phosphate. This phase starts to become distinguishable in the diffraction pattern of samples calcined at 800°C.

Similar results have been reported by several investigators, which suggest that the formation temperature of β-tricalcium phosphate is between 750°C - 800°C [21].

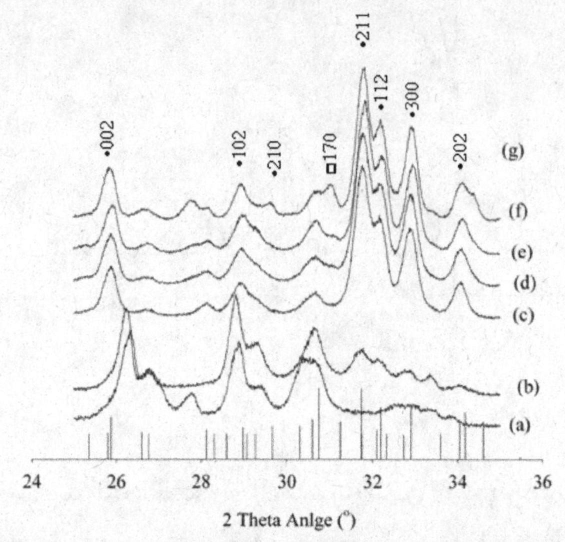

Figure 4. X-Ray diffraction pattern of the gel calcined at different temperatures for 0.5hr (a) 300°C, (b) 400°C, (c) 500°C, (d) 600°C (e) 700°C (f) 800°C. The primary diffraction peaks for hydroxyapatite (●) and β-tricalcium phosphate (□) are indicated.

The crystal size was calculated based on the X-ray diffraction pattern according to the Scherrer equation [26] to further investigate the influence of the calcination temperature.

$$B = \frac{0.9\lambda}{t\cos\theta} \quad (1)$$

where B equals to FWHM (full width at half maximum) of the broadened diffraction line on the 2θ scale, which, in this investigation, is the reflection of (002) and (211) planes, λ is the wavelength for Cu-Kα (λ =0.15418 nm) and t is the diameter of the crystallites. The average value was taken as the final value of the crystal size and was typically between 20 and 40 nm. The crystal size data as a function of calcination temperature are plotted in Figure 5.

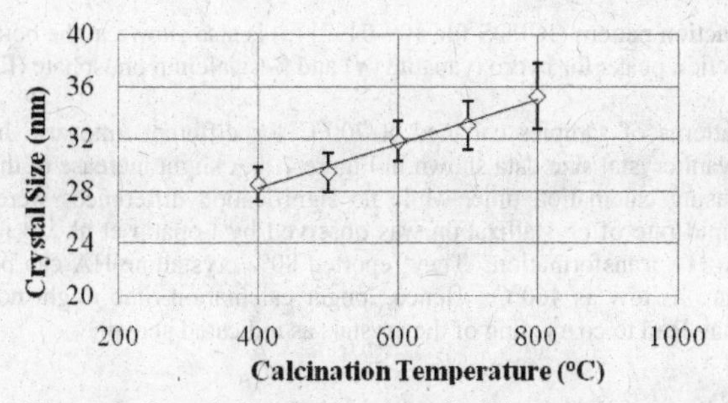

Figure 5. Crystal size as a function of calcination temperature.

It is observed that the crystal size increases linearly with increasing calcination temperature. This result is consistent with the data of Liu et al. [18] who studied line-breadth of the (002) peak at 2θ=25.83° of the apatitic powder for gels calcined at different temperatures. The half-intensity width of the (002) peak decreased gradually with a corresponding improvement in the sharpness of the major peaks as the temperature increased from 350°C to 800°C. This observation indicates an increase in crystallite size or improved crystal perfection of the calcined HA powders. On the other hand, an increase in the crystal size might cause coarsening of the final powder. Thus, high temperature calcination may favor the formation of a secondary phase as well as an increase of the crystal size, which is not desirable in some of applications.

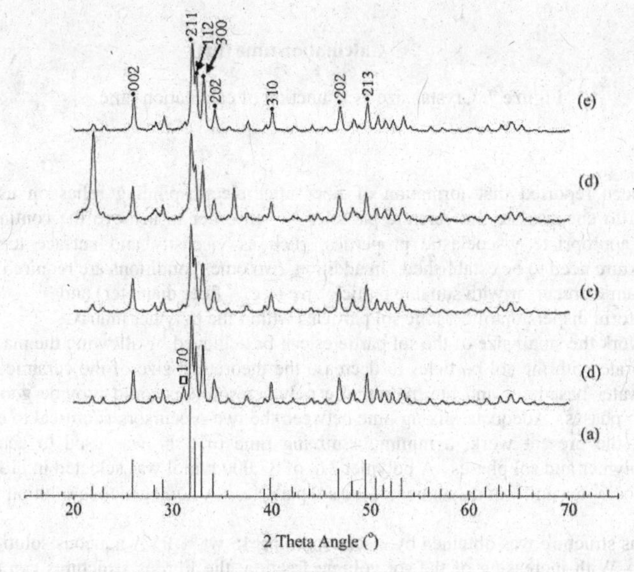

Figure 6. X-ray data of 700°C calcined 70°C 24 hr Samples: (b) 700°C calcining for 2hr; (c) 700°C calcining for 3hr; (d) 700°C calcining for 4hr; (e) 700°C calcining for 5hr. The standard

hydroxyapatite diffraction pattern (JCPDS file #09-0432) (a) is also shown at the bottom of the figure. The primary diffraction peaks for hydroxyapatite (•) and β-tricalcium phosphate (□) are indicated.

The XRD patterns of samples calcined at 700°C for different time are shown in Figure 6 followed by the relevant crystal size data shown in Figure 7. A slight increase of the crystal size was observed with increasing calcination time, while no signification differences were observed in the XRD patterns. A rapid rate of crystallization was observed by Lopatin et al. [22] for the amorphous calcium phosphate to HA transformation. They reported 80% crystalline HA can be obtained within 200 s at a temperature as low as 460°C. Hence, longer calcination time might not favor increased crystallization, but may lead to coarsening of the crystals as indicated above.

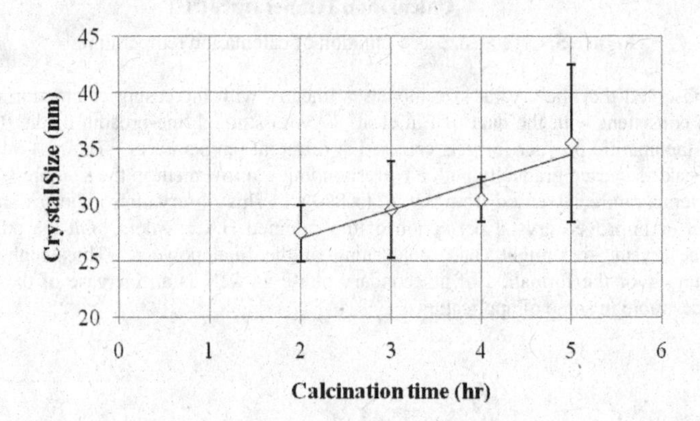

Figure 7. Crystal size as a function of calcination time.

Electrospinning

It has been reported that formation of fiber after electrospinning relies on using polymer macromolecules to encapsulate the ceramic particles [23]. In order to achieve the continuous fibrous structures, the appropriate viscoelastic properties, such as viscosity and surface tension, of the polymer/sol mixture need to be established. In addition, two other conditions are required:

1. A ceramic precursor with suitable particle size (e.g. < fiber diameter) and
2. A uniform dispersion of ceramic sol particles within the polymer matrix.

In this work the small size of the sol particles can be achieved by allowing the macromolecules to fully incorporate with the sol particles to decrease the theoretical size of the ceramic particles. In this case, the water based sol and the hydrophilic polymer solution could provide good miscibility between the two phases. Adequate mixing time between the two precursors is critical to ensure proper miscibility. In the present work, a minimum mixing time of 8 hr was used to achieve a good dispersion of polymer and sol phases. A polymer M_W of 67,000 g/mol was selected in order to achieve a moderate viscosity for uniform dispersion of the sol particles. A homogeneous solution was obtained after mixing.

A fibrous structure was obtained by electrospinning 15 wt.% PVA aqueous solution, as shown in Figure 8 (a). With increasing of the sol volume fraction, the fibrous structures can remain up to 50% of sol. The fibrous structures are shown in Figures 8 (b) and (c).

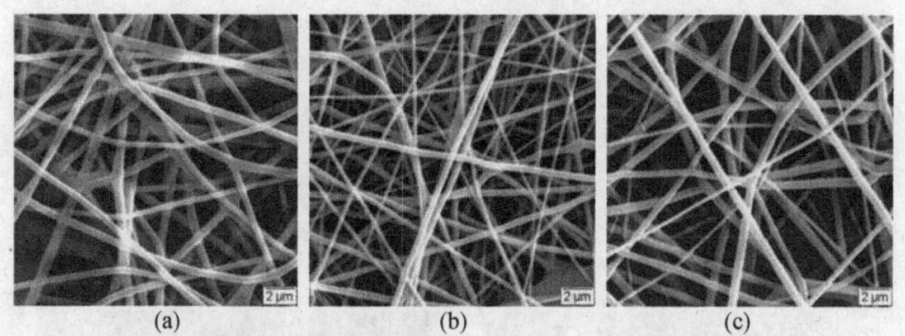

(a) (b) (c)

Figure 8 Photographs showing the as-spun structures for different sol volume fractions. (a) 0% (PVA aqueous solution); (b) 16.7%; (c) 20%; (PVA M_W = 67,500 g/mol).

Calcination of electrospun Scaffolds

The dried electrospun hybrid samples were calcined at 600°C for 6 hr. ThepPolymer can be totally eliminated upon calcinations, while the as-spun fibrous morphologies were generally retained after calcination. Figure 9 shows the interconnected structure obtained after calcination. Upon calcination of the dried sample, a consolidation reaction occurs. The sol particles tend to sinter together to minimize the surface area. The formation of these highly interconnected structures can be explained as follows (Figure 10). A fiber mat with a high fiber density forms after electrospinning the polymer-sol mixture. As a result, there are a large number of intersections between fibers, at which sol particles may sinter together and form junctions. The remaining sol particles may be stretched towards the ends and connected at the junctions (Figure 10). Note that a more open structure with thinner fibers between the junctions is obtained with sol volume fraction of 20% than with a sol volume fraction of 16.7% because of the initial electrospun structure. As discussed before, the thinner fibers are obtained upon electrospinning as the sol content increases from 16.7% to 20%.

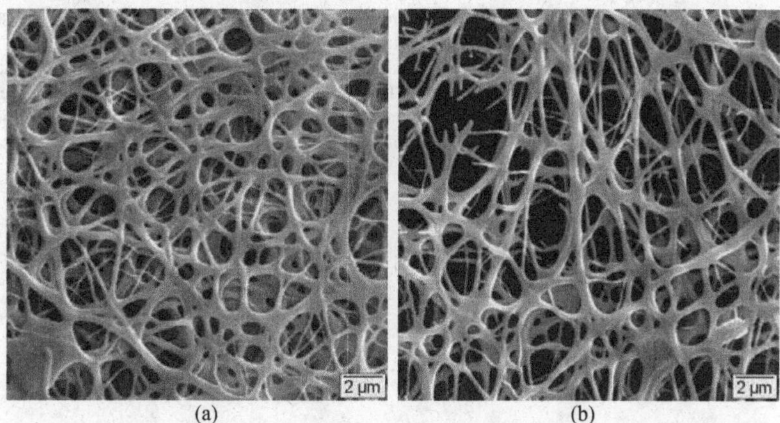

(a) (b)

Figure 9 Photographs showing highly interconnected structures after calcining at 600°C for 6 hr for sol volume fractions of (a) 16.7% and (b) 20% (PVA M_W =67,500 g/mol).

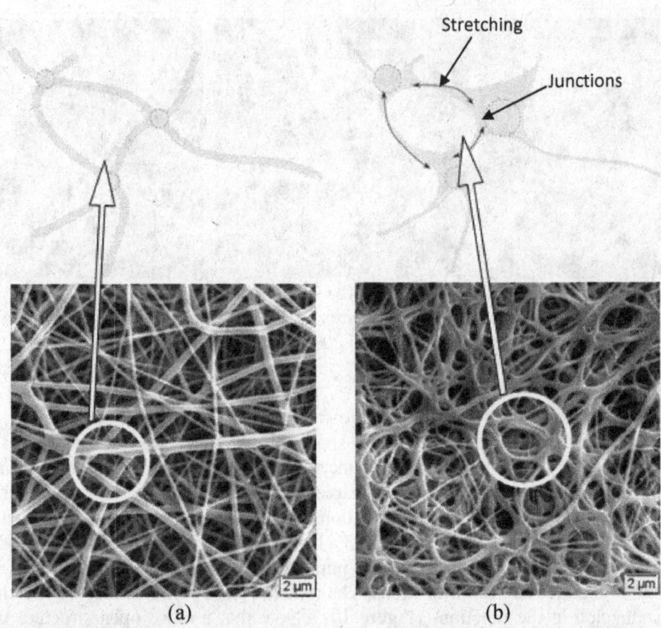

(a) (b)

Figure 10 SEM photographs are shown before (a) and after (b) calcining. The highlighted regions are shown in detail to outline the formation of interconnected porous structures for the two cases. The stretching of the fibers between junctions is illustrated (PVA M_W = 67,500 g/mol).

The diffraction patterns with different sol volume fraction are shown in Figure 11. It can be seen that the sol volume fraction did not have a significant effect on the crystal structure. Crystal sizes were measured to be on the order of 10-15 nm. This value is much smaller than the value obtained with powders produced from pure sol (30 nm). Thus it appears that electrospinning does seem to reduce the crystal size in the ceramic. Several investigators have reported that the crystal size in the polymer decreases upon electrospinning [25]. A similar effect may be observed in the calcined ceramic.

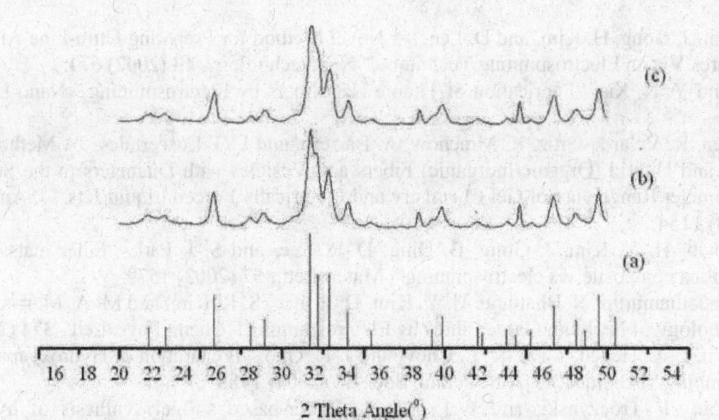

Figure 11 XRD pattern for samples calcined at 600°C for 6 hr. (b) Sol volume fraction = 16.7%, and (c) Sol volume fraction = 20. The JCPDS file (#09-0432) for hydroxyapatite is also shown in the bottom of the figure (a).

CONCLUSIONS

Fibers and highly porous architectures of hydroxyapatite could be produced after calcination of an electrospun PVA/calcium phosphate based sol precursor. Complete fibrous structures were obtained after electrospinning and calcination. The aging and calcination conditions significantly affect the formation of the final ceramic phase. XRD analysis indicated hydroxyapatite to be the dominant inorganic phase remaining after calcination with an average crystal size about 30 nm. Electrospinning of the mixture reduces the crystal size in the hydroxyapatite.

REFERENCE

[1] L. L. Hench, "Sol-gel materials for bioceramic applications." Curr. Opin. Solid State Mater. Sci., **2** (1997) 604.

[2] L. L. Hench, "Bioceramics." J Am. Ceram. Soc., **81** (1998) 1705.

[3] L. L. Hench, "Bioceramics, a clinical success." Am. Ceram. Soc. Bull., **77** (1998) 67.

[4] L. L.Hench, "Challenges for bioceramics in the 21st century." Am. Ceram. Soc. Bull., **84** (2005) 5.

[5] L. L. Hench, D. L. Wheeler, D. C. Greenspan, "Molecular control of bioactivity in sol-gel glasses." J Sol Gel Sci Technol, **13** (1998) 245.

[6] Paschalis E. P., Q. Zhao, B. E. Tucker, S. Mukhopahayay, J. A. Bearcroft, N. B. Beals, M. Spector, and G. H. Nancollas. "Degradation potential of plasma-sprayed hydroxyapatite-coated titanium implants. J. Biomed. Mater. Res., **29** (1995) 1499.

[7] N. Yamasaki, T. Kai, M. Nishioka, K. Yanagisawa, and K. Ioku, "Porous hydroxyapatite ceramics prepared by hydrothermal hotpressing." J. Mater. Sci., **9** **(**1990) 1150.

[8] A. Slosarczyk, E. Stobierska, Z. Paszkiewicz, and M. Gawlick, "Calcium Phosphate Materials Prepared from Precipitates with Various Calcium:Phosphorus Molar Ratios." J. Am. Ceram. Soc., **79** (1996) 2539.

[9] M. Yoshimura, H. Suda, K. Okamoto, and K. Ioku, "Hydrothermal synthesis of biocompatible whiskers." J. Mater. Sci., **29** (1994) 3399.

[10] L.C. Klein, "Sol-Gel Technology for Thin Films, Fibers, Preforms, Electronics and Specialty Shapes." 1988 William Andrew Publishing/Noyes

[11] A. Nazeri, E. Bescher, and J. D. Mackenzie, "Ceramics composites by the Sol-Gel Methods: A review." Ceram. Eng. Sci. Proc., **14** (1993) 1.

12. H. Q. Dai, J. Gong, H. Kim, and D. Lee, "A Novel Method for Preparing Ultra-Fine Alumina-Borate Oxide Fibres Via an Electrospinning Technique." Nanotechnology, **13** (2002) 674.
13. D. Li, and Y. N. Xia, "Fabrication of Titania Nanofibers by Electrospinning." Nano Lett., **3** (2003) 555.
14. G. Larsen, R. Velarde-Ortiz, K. Minchow, A. Barrero, and I. G. Loscertales, "A Method for Making Inorganic and Hybrid (Organic/Inorganic) Fibers and Vesicles with Diameters in the Submicrometer and Micrometer Range via Sol-Gel Chemistry and Electrically Forced Liquid Jets." J. Am. Chem. Soc., **125** (2003) 1154.
15. C. L. Shao, H. Y. Kim, J. Gong, B. Ding, D.-R. Lee, and S.-J. Park, "Fiber mats of poly(vinyl alcohol)/silica composite via electrospinning", Mater. Lett., **57** (2003) 1579.
16. P. Viswanathamurthi, N. Bhattarai, H. Y. Kim, D. R. Lee, S. R. Kim, and M. A. Morris, "Preparation and Morphology of Niobium Oxide Fibres by Electrospinning." Chem. Phys. Lett., **374** (2003) 79.
17. Y. Q. Wu, L. L. Hench, J. Du, K. L. Choy, and J. K. Guo, "Preparation of Hydroxyapatite Fibers by Electrospinning Technique." J. Am. Ceram. Soc., **87** (2004) 1988.
18. D.-M. Liu, T. Troczynski, and W.J. Tseng, "Water-based sol-gel synthesis of hydroxyapatite: process development." Biomaterials, **22** (2001) 1721.
19. D.-M. Liu, T. Troczynski, and W.J. Tseng, "Aging effect on the phase evolution of water-based sol-gel hydroxyapatite." Biomaterials, **23** (2002)1227.
20. C. S.Chai, B. Ben-Nissan, S. Pyke, and L. Evans, "Sol-gel derived hydroxyapatite coatings for biomedical applications." Mater. Manuf. Process, **10** (1995) 205.
21. C. S. Chai, K. A. Gross, and B. Ben-Nissan, "Critical ageing of hydroxyapatite sol-gel solutions." Biomaterials, **19** (1998) 2291.
22. C. M. Lopatin, V. B. Pizziconi, and T. L.Alford. "Crystallization kinetics of sol-gel derived hydroxyapatite thin films." J. Mater. Sci. - Mater. Med., **12**(2001) 767.
23. W. Sigmund, J. Yuh, H. Park, V. Maneeratana, G. Pyrgiotakis, A. Daga, J. Taylor, and J. C. Nino, "Processing and Structure Relationships in Electrospinning of Ceramic Fiber Systems." J. Am. Ceram. Soc., **89** (2006) 395.
24. L. H. Sperling, Introd. Phys. Polym. Sci. 2nd Ed. NY: Wiley; 1992.
25. J. Yuh, J. C. Nino, and W. Sigmund, "Synthesis of Barium Titanate ($BaTiO_3$) Nanofibers via Electrospinning." *Mater. Lett.,* **59** (2005) 3645.
26. B.D. Cullity, and S.R.Stock, "Elements of X-Ray diffraction." 3rd edition, New Jersy: Prentice-Hall, Inc.

Millimeter Wave Properties of Titania Photonic Crystals with Diamond Structures Fabricated by Using Micro-stereolithography

M. Kaneko, S. Kirihara
Joining and Welding Research Institute, Osaka University
11-1 Mihogaoka, Ibaraki, Osaka 567-0047, Japan

ABSTRACT

Three dimensional micro photonic crystals with a diamond structure made of a dense titania were fabricated, and photonic band gap properties in the millimeter waveguides were investigated. Acrylic diamond lattice structures with titania particles dispersion at 40 vol. % were fabricated by using micro-stereolithography. The forming accuracy was 10 μm. After sintering process, the titania diamond lattice structures were obtained. The relative density reached 96 %. Millimeter wave transmittance properties were measured by using a network analyzer and a W-band millimeter waveguide. In the transmission spectra for the Γ-X <100> direction, a forbidden band was observed from 90 to 110 GHz. The frequency range of the band gap well agreed with calculated results by plane wave expansion (PWE) method. Additionally, simulated results by transmission line modeling (TLM) method indicated that a localized mode can be obtained by introducing a plane defect between twinned diamond lattice structures.

INTRODUCTION

Photonic crystals are periodic arrangement structures of dielectric medium and form photonic band gaps which can prohibit propagations of electromagnetic waves by Bragg Reflection [1, 2]. The band gap properties can be controlled arbitrarily by changing the structures of photonic crystals. For example, by introducing artificial plane defects of air gaps in the photonic crystals, resonant modes of transmission peaks can be obtained in the photonic band gaps. At these frequencies, the electromagnetic waves were permitted to transmit corresponding to the defect length. By using these interesting features, the photonic crystals can be applied to resonators, wavelength filters, directional antennas and so on [3, 4]. In particular, three dimensional photonic crystals with diamond structures are regarded as ideal dielectric patterns, since they have complete photonic band gaps which can prohibit the incident electromagnetic waves in all directions [5]. However, it has been difficult to fabricate three dimensional dielectric objects with complex structures. In our previous study, we have successfully fabricated the photonic crystals with the diamond structures composed of ceramic particles dispersed photosensitive resin by using the stereolithography method of a computer aided design and manufacturing (CAD/CAM) process [6, 7]. Future applications using millimeter waves along 100 GHz are expected in various areas, such as wireless communication systems, intelligent traffic systems

(ITS), environmental analysis through remote sensing methods and so on. In order to control the millimeter waves in the higher frequency range, we have to develop a new technique to fabricate the micrometer order three dimensional objects. Recently, our research group succeeded to create the micro photonic crystals composed of dielectric material such as alumina, silica and titania dispersed resin composites by micro-stereolithography [8, 9]. These photonic crystals can control the electromagnetic waves in the GHz and THz frequency ranges. However, the structural designs of the photonic crystals to modify the band gap profiles were not really mentioned in these reports. Especially, it is important to investigate, how many numbers of the unit cells should be necessary in the vertical directions to the incident electromagnetic waves to obtain the wider photonic band gaps. In this research, we attempted to fabricate the diamond photonic crystals which can control the millimeter wave transmissions in the waveguides of metal cavities. In the beginning, we selected the dielectric materials and the designed structural dimensions of the diamond lattices to open the forbidden gaps within the transmission spectrum on W-band from 75 to 110 GHz. Subsequently, we fabricated the titania micro photonic crystals with the diamond lattice structures through the micro-stereolithography of the CAD/CAM process and the conventional sintering processes. Their Measured millimeter wave properties were compared with an electromagnetic band diagram simulated theoretically by solving Maxwell equations. To discuss the real application of the titania photonic crystals to the wave select filters, the millimeter wave resonations in the plane defects introduced into the diamond lattice structures were simulated through finite element method.

DESIGN AND CALCULATION

Generally, the band gap frequency can be modulated by controlling the lattice constant, the dielectric constant and the volume fraction of the dielectric materials. A unit cell of the diamond structure was designed as shown in Fig. 1 (a) by using a 3D CAD software (Toyota Caelum Co. Ltd., Think Design, and Ver. 9.0). The aspect ratio of the dielectric lattice was 1.5. The volume fraction of the dielectric bodies was 33 %. An electromagnetic band diagram of the designed diamond structure was calculated along the symmetry lines in the Brillouin zone by means of plane wave expansion (PWE) method as shown in Fig. 1 (b) [10]. The dielectric constant of 100 was substituted for the material parameter of the titania lattice. In the calculation, plane waves of 124 in number propagated into the imaginarily limitless periodic structure of the dielectric lattice. Through the diamond lattice structure, the perfect photonic band gap was opened for all crystal directions. Fig. 2 (a) shows the perfect band gap frequencies through the diamond structures composed of the titania with the various lattice constants. The solid and dotted lines indicate the higher and lower edges of the band gap frequencies, respectively. The diamond structure with the longer lattice constant exhibits the forbidden band at the lower frequency range. In order to control the millimeter wave propagation effectively through the photonic crystal, whole dimension of the lattice structure should be reduced to introduce the enough unit cells into the waveguide of metal cavities for the electromagnetic wave diffraction. Moreover, the photonic band gap should be opened from 75 to 110 GHz, in which the standard transmission mode of the millimeter wave was formed through the W-band waveguide. As shown in Fig. 2 (a), the gray region indicates the W-band frequency, and the ideal lattice constant was estimated to be 0.72 mm for the band gap formation in the waveguide. The whole crystal structure fitting to the W-band waveguide of $1.27 \times 2.54\ mm^2$ in cross sectional dimensions was designed as shown in Fig. 2 (b). The diamond lattice was consisted of $1.75 \times 3.5 \times 4$ unit cells in numbers.

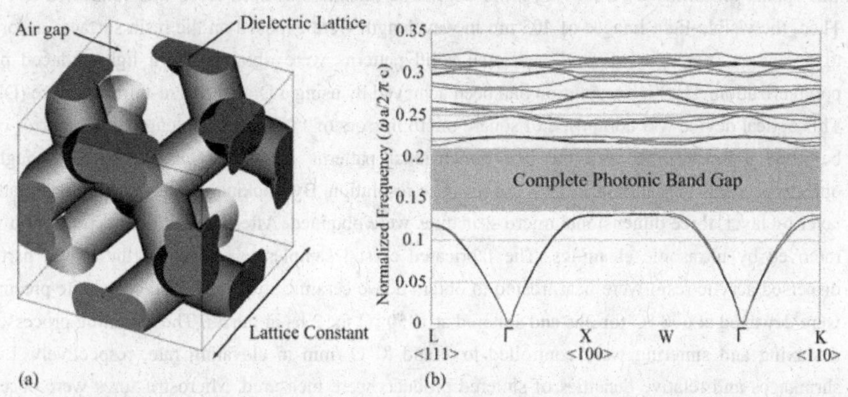

Fig. 1 A designed computer graphic model of a unit cell of a photonic crystal with a diamond structure (a), and an electromagnetic band diagram calculated by using plane wave expansion (PWE) method (b). The dielectric constant and the volume fraction of the lattice were 100 and 33 %, respectively.

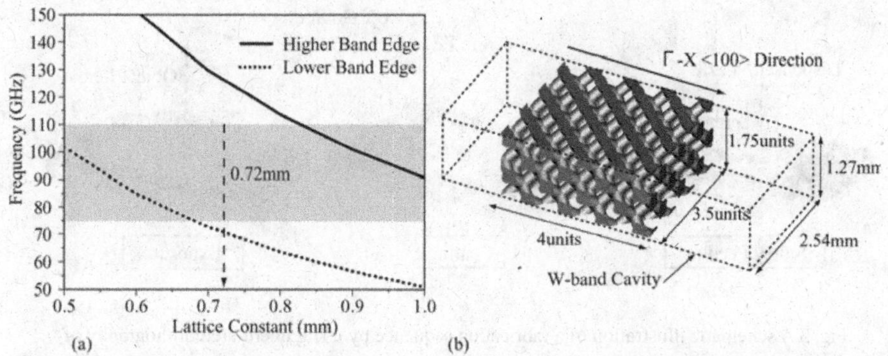

Fig. 2 Frequencies variations of perfect photonic band gaps of diamond structures composed of titania according to lattice constants calculated by using PWE method (a), and a designed graphic model of a crystal sample fitted to a metal cavity in a W-band waveguide (b).

EXPEIMENTAL PROCEDURE

The diamond photonic crystals were fabricated by using the micro-stereolithography. The designed model was converted into the stereolithography (STL) files and sliced into a series of two dimensional layers. These data were transferred to the fabrication system (D-MEC Co. Ltd., SIC-1000). Fig. 4 shows a schematic illustration of the fabrication process by the stereolithography. Nanometer sized titania particles of 270 nm in diameter were dispersed into the photo sensitive acrylic resin at 40 vol. %. The mixed slurry was squeezed on a working stage from a dispenser nozzle. This material paste was spread uniformly by a moving knife edge. The thickness of each layer was controlled to 5 μm. Then, the visible-light images of 405 nm in wavelength were exposed on the resin surfaces according to computer operations. Two dimensional solid patterns were obtained by a light induced photo polymerization. The high resolution had been achieved by using a Digital Micro-mirror Device (DMD). This optical device was composed of square micro mirrors of 14 μm in edge length. Each mirror could be tilted independently, and the two dimensional patterns are dynamically exposed through an objective lens as bitmap images of 2 μm in space resolution. By stacking these cross sectional patterns layer by layer, three dimensional micro structures were obtained. After fabricating, uncured resin were removed by ultrasonic cleanings. The fabricated crystal samples composed of the titania particles dispersed acrylic resin were heat treated to obtain dense ceramic structures. The composite precursors were dewaxed at 600 °C for 2hs and sintered at 1350 °C for 2 hs in the air. These heating processes on dewaxing and sintering were controlled to 1 and 8 °C /min in elevating rate, respectively. Linear shrinkages and relative densities of sintered products were measured. Microstructures were observed by using scanning electron microscopes (SEM). A bulk sample of the sintered titania was also fabricated to measure the dielectric constant.

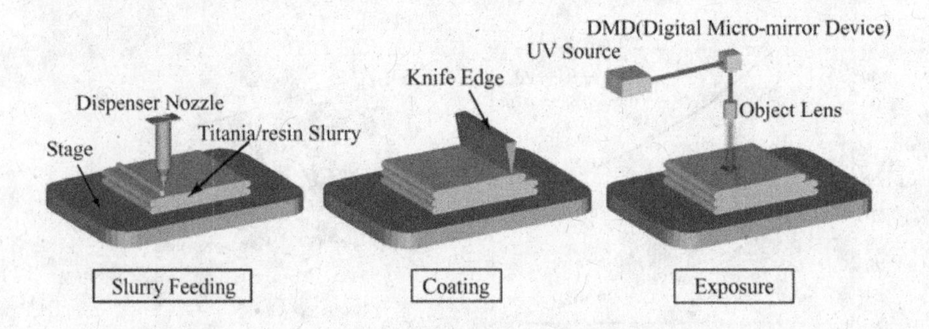

Fig. 3 A schematic illustration of a fabrication sequence by using micro-stereolithography of a computer aided designing and manufacturing (CAD/CAM) process.

The Attenuations of the millimeter-wave transmissions through photonic crystals were measured in the frequency range from 75 to 110 GHz by using a W-band millimeter waveguide and a network analyzer (Agilent Technology Co. Ltd., E8362B). The millimeter wave was transmitted trough the crystal samples for the Γ-X <100> direction. The dielectric constant and loss of the titania bulk sample was measured by a using terahertz time domain spectrometer (Advanced Infrared Spectroscopy Co. Ltd., J-Spec 2001 spc). Subsequently, the diamond photonic crystals with or without a plane defect were designed successfully, and the millimeter wave propagation behaviors through the lattice structure in the W-band waveguide were simulated by using transmission line modeling (TLM) method (Flomerics Group, Micro-stripes Ver. 7.5) [11].

RESULTS AND DISCUSSION

Figure 4 (a) shows the top view of the titania dispersed acryl photonic crystal with a diamond structure fabricated by the micro-stereolithography. The lattice constant of the formed structure was 0.96 mm. Dimensional differences between the designed model and formed sample were less than 10 μm. Fig. 4 (b) shows the sintered titania photonic crystal. Linear shrinkage through the sintering was about 25 %. Then, the lattice constant of the sintered sample reduced to 0.72 mm. The obtained samples could be inserted closely into the metal cavity of the W-band millimeter waveguide. Fig. 4 (c) shows a SEM image of the sintered sample. Cracks or pores were not observed in the microstructures of the sintered samples. Relative density of the sample reached 96 %. The dielectric constant of the sintered titania was measured as 100 along 100 GHz.

Fig. 4 A diamond type acryl lattice with titania particles dispersions formed by the micro-stereolithography (a), a sintered titania photonic crystals and a SEM micrograph of the dielectric lattice (c).

The millimeter wave transmission spectra through photonic crystals for the Γ-X <100> direction are shown in Fig. 5. These transmission intensities were measured by using the network analyzer and calculated by the TLM method. The solid and dotted lines show the measured and simulated transmission spectra, respectively. The measured results are nearly coincided with the simulated one. The forbidden band prohibiting the millimeter wave transmission is formed from 90 to 110 GHz. This frequency range was included in the perfect photonic band gap, which was calculated by PWE method, to diffract the electromagnetic waves effectively for all crystal directions.

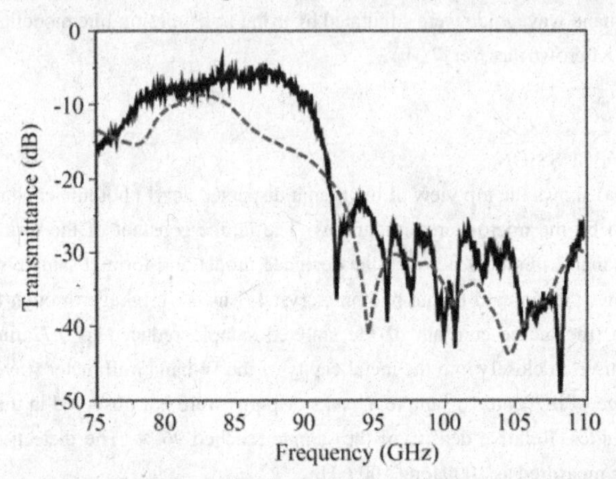

Fig. 5 Transmission spectra through the titania photonic crystal for the Γ-X <100> direction. The solid lines show the measured transmittance in the W-band waveguide by using a network analyzer, and the dotted lines show the simulated one through transmission lime modeling (TLM) method.

The millimeter wave localization in a twinned diamond structure with a plane defect was investigated by using the TLM simulation. In the photonic crystal with a twinned diamond structure, the plane defect was formed as an interface between the mirror symmetric titania lattices, as shown in Fig. 6 (a). The twinned interfaces parallel to (100) lattice plane was sandwiched between the two titania lattice components with one period structure. In the simulated transmission spectrum obtained by the TLM method, one resonant peak was formed at 99 GHz in the photonic band gap. At this frequency, the electromagnetic wavelength is equal to the lattice constant in the dielectric crystal structure. Fig. 6 (b) shows a cross sectional distribution image of the electric field intensity in the twinned lattice structure at the resonant frequency. The black and white areas indicate the electric field intensity is high and low, respectively, and the electromagnetic waves propagate for the right direction. The incident electromagnetic wave is resonated and localized in the defect interface through the multiple reflections between the mirror symmetric diffraction lattices. This amplified electromagnetic wave can transmit the opposite side of the photonic crystal and form the transmission peak in the photonic band gap.

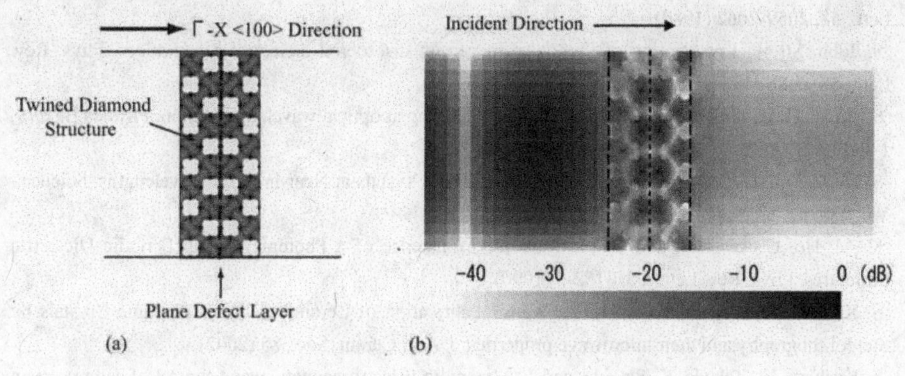

Fig. 6 A schematic model of a twinned diamond structure with a plane defect between mirror symmetric dielectric lattices (a), and a cross sectional distribution image of electric field intensities in vicinity area to the defect interface at the resonant frequency simulated by using the TLM method.

CONCLUSION

Titania photonic crystals composed of fine dielectric lattices with diamond structures were developed to control millimeter waves. These lattice structures were designed to realize photonic band gaps formations in a metal cavity of millimeter waveguide at W-band from 75 to 110 GHz in frequency range. Acryl lattices with titania particles dispersions at 40 % in volume contents were successfully fabricated by using micro-stereolithography of a CAD/CAM process. The forming tolerance was less than 10 μm. Titania photonic crystals with the diamond structures were obtained through dewaxing and sintering processes. The relative density reached 96 %. The obtained titania photonic crystals of 0.72 mm in lattice constant and 1.75 × 3.5 × 4 unit cells in numbers could be inserted closely into the millimeter waveguide of 1.27 × 2.54 mm^2 in cross sectional dimensions. The photonic band gap could be formed form 90 to 110 GHz. The measured result had good agreement with the theoretically calculated one by using PWE and TLM method. A twinned diamond lattice structure with a plane defect interface formed a resonance peaks in the band gap through theoretical simulations. It is expected that the titania photonic crystals can be applied to control millimeter waves in the metal cavities as micro filtering devices and so on.

REFERENCES

[1]E. Yablonovitch, Inhibited Spontaneous Emission in Solid-State Physics and Electronics, Phys. Rev. Lett., 58, 2059-2062 (1987).

[2]S. John, Strong Localization of Photonics in certain disordered dielectric superlattices, Phys. Rev. Lett., 58, 2486 (1987).

[3]S. Noda, Three-dimensional photonic crystals operating at optical wavelength region, Physica B., 279, 142-149 (2000).

[4]S. Noda, Full Three-Dimensional Photonic Bandgap Crystals at Near-Infrared Wavelengths, Science., Vol. 289, no. 5479, pp. 604 – 606 (2000).

[5]K. M. Ho, C. T. Chan, and C. M. Soukoulis, Existence of a Photonic Gap in Periodic Dielectric Structures, Phys. Rev. Lett., 65, 3152-5 (1990).

[6]S. Kirihara, Y. Miyamoto, and K. Kajiyama, Fabrication of Ceramic-Polymer Photonic Crystals by Stereolithography and their microwave properties, J. Am. Ceram. Soc., 85 (2002).

[7]S. Kirihara, M. Takeda, K. Sakoda, and Y. Miyamoto, Electromagnetic wave control of ceramic/resin photonic crystals with diamond structure, Science and Technology of Advanced Materials 5., 225-230 (2004).

[8]W. Chen, S. Kirihara, and Y. Miyamoto, Fabrication and Measurement of Micro Three-dimensional Photonic Crystals of SiO_2 Ceramic for Terahertz Wave Applications, J. Am. Ceram. Soc., 90 (2007).

[9]W. Chen, S. Kirihara, and Y. Miyamoto, Fabrication of 3D Micro Photonic Crystals of resin Incorporating TiO_2 Particles and Their Terahertz Wave Properties, J. Am. Ceram. Soc., 90, 92-96 (2007).

[10]K. M. Ho, C. T. Chan, and C. M. Soukoulis, Existence of a Photonic Gap in Periodic Dielectric Structures, Phys. Rev. Lett., 65, 3152-5 (1990).

[11]J. A. Morente, G. J. Molina-Cuberos, J. A. Porti, K. Schwingenschuh, and B. P. Besser, A study of the propagation of electromagnetic waves in Titan's atmosphere with the TLM numerical method, Icarus, 162, 374-384 (2003).

SINTERING KINETIC STUDY OF 2Y-TZP/AL_2O_3 COMPOSITE DURING INITIAL STAGE OF SINTERING

Abhijit Ghosh*, Soumyajit Koley, Ashok Kumar Sahu, Ashok Kumar Suri
Materials Group, Bhabha Atomic Research Centre, Mumbai, India

ABSTRACT

Initial stage of sintering of nano crystalline 2 mol% yttria doped tetragonal zirconia polycrystal (2Y-TZP) with 2wt% alumina (2Y-TZP/Al_2O_3) was analyzed using different nonisothermal and pseudo-isothermal sintering techniques. Sintering behaviour of the composite material during constant rate of heating (CRH) was found to be dependent on calcination temperature. Kinetics of sintering was studied using pseudo-isothermal sintering method, i.e. stepwise isothermal densification (SID) technique. Although this method was expected to produce both 'n' and 'Q' simultaneously from a single experiment, the result obtained in the present experiment was not reliable, especially the 'Q' value obtained was abnormally large as compared to the value expected for a nano crystalline material. However, the same SID experiment was found to be useful to measure the activation energy by Dorn method. In the present investigation, activation energy of sintering was evaluated at different temperature regions using Dorn method. The modified Johnson equation was used for the direct determination of 'n' value. Incorporation of initial shrinkage correction during SID experiment yielded reliable 'Q' as well as 'n' values. Most interesting finding of the present study was that the sintering mechanism, as a function of calcinations temperature, was found to change from grain boundary to volume diffusion mechanism.

Key words: zirconia, alumina, sintering, dilatometry, activation energy

INTRODUCTION

Various sintering rate equations have been used to describe the kinetics of the initial stage of sintering. [1-7] Isothermal sintering experiments have provided most of the available sintering data. [7-9] However, reproducibility in experimental data is poor in isothermal sintering of ceramics which have poor thermal conductivity. In ceramics, initiation of shrinkage takes place before reaching a steady state temperature. To circumvent this, sintering kinetics in the ceramic systems have also been studied under constant rate of heating (CRH).[10] Traditionally there is a belief that it is not possible to determine simultaneously both mechanism of sintering and activation energy i.e. n and Q respectively from single CRH experiment.[2, 11] Lahiri et al[12] tried to utilize single CRH experiment to determine the activation energy directly. They used the relation proposed by Wang and Raj[3] and reformulated it in the following form (Eq.1):

$$\ln[Ta\frac{dy}{dt}] = -\frac{Q}{RT} + \ln k_o + \ln f(y) - \ln G^{\alpha k} \text{ -------------------------------- (1)}$$

where, y= relative shrinkage (=$\nabla L/L_o$, where ∇L is shrinkage = L_o-L_t, L_o=initial length and L_t= length at time t), dy/dt is the rate of shrinkage, k_o is a constant, Q is the activation energy, T is the temperature (K), "a" is constant heating rate, $f(y)=ny^{1-\frac{1}{n}}$, G is the crystallite size and "α" and "k" are two constants. Surprisingly, to determine activation energy directly, they had chosen to plot left hand side of Eq.1 as a function of "1/T", ignoring "f(y) term completely. Since "f(y)' is a time and temperature dependent variable, this approach is bound to lead towards erroneous result. Moreover,

* author for correspondence, email: abhijitghosh72@indiatimes.com; phone: 912225590496; Fax: 912225505151. Address: ACS, MPD, Materials Group, Mod Lab, BARC, Mumbai, India, 400085.

they used modified Dorn method (described in experimental section) for calculating activation energy from CRH experiment, which is improper. [11, 13] Perez-Maqueda et al[5] calculated both 'n' and 'Q' simultaneously from a single CRH experiment by modifying the existing sintering rate equation to the following form (Eq.2):

$$\frac{d(\ln\frac{dy}{dT})}{d(\ln y)} - a[\frac{d(\ln T)}{d(\ln y)}] = -\frac{Q}{R}[\frac{d(\frac{1}{T})}{d(\ln y)}] + 1 - \frac{m}{2} \text{ ---------- (2)}$$

where $m=\frac{2}{n}$, and 'a' is an exponent. Considering presence of either grain boundary diffusion (GBD) or volume diffusion (VD) as the dominant sintering mechanism, value of 'a' becomes -1. According to them, a plot of left hand side of Eq.2 against $[\frac{d(\frac{1}{T})}{d(\ln y)}]$ would lead to a straight line whose slope would give $\left(-\frac{Q}{R}\right)$ and intercept $(1-\frac{m}{2})$. Eq.2 can also be written in the following form (Eq.3):

$$y\frac{\frac{d^2y}{dT^2}}{(\frac{dy}{dT})^2} - a\frac{y}{T\frac{dy}{dT}} = \frac{Q}{R}[\frac{y}{T^2\frac{dy}{dT}}] + 1 - \frac{m}{2} \text{ ---------- (3)}$$

It indicates that Perez-Maqueda et al[5] used a second order, second derivative equation, which may not give a linear relation. Laine et al[14] used Eq.2 to study the sintering kinetic of nano alumina system. The "n" value obtained by them from this relation was 2.0, which does not fit to any of the known sintering mechanisms.[2]

From the work of Wang and Raj[3], and Matsui et al.,[6] it is evident that minimum of four CRH experiments are required to determine 'Q'. In this respect a quasi-isothermal technique[15] like SID (stepwise isothermal Dilatometry) has proved to be very useful. In this, the following concept was used[16]: the sample was heated at a constant heating rate until the rate of shrinkage exceeded a preset limit, at which point heating was stopped and the shrinkage took place in an isothermal condition until the rate of shrinkage became smaller than the limit where the heating was resumed. Initially Ali and Sorensen[17] used the following relation (Eq.4 and 5) to determine both 'n' and 'Q':

$$\frac{dy}{dt} = nK(T)y^{1-\frac{1}{n}} \text{ ---------- (4)}$$

$$\text{where, } K(T) = K_o \exp(-\frac{Q}{RT}) \text{ ---------- (5)}$$

Plot of $\ln(\frac{dy}{dt})$ vs. ln(y) from Eq.4 for different isothermal steps have been utilized for determining the mechanism of sintering (slope) and ln[K(T)] (from intercept). Activation energy can be calculated from the slope of the plot of ln[K(T)] vs. 1/T. Subsequently, to describe dynamic relative volumetric shrinkage, Eq.4 was modified to the following form[18] (Eq.6):

$$\frac{V_O - V_t}{V_t - V_f} = [K(T)(t - t_O)]^n \text{ ---------- (6)}$$

where, t_o is the time of initiation of isothermal step, V_O, V_t and V_f are the initial volume, volume at time t and fully dense volume of the specimen respectively. Assuming shrinkage in a polycrystalline material as isotropic in nature, several researchers[19-22] modified Eq.6 to the following form (Eq.7):

$$\frac{dY}{dt} = nK(T)Y(1-Y)\left(\frac{1-Y}{Y}\right)^{\frac{1}{n}} \text{ ---------- (7)}$$

where, $Y=\frac{V_O-V_t}{V_O-V_f}=\frac{L_O^3-L_t^3}{L_O^3-L_f^3}$; L_t and L_f are the length at time t and of fully dense sample respectively. A plot of $\ln\left[\frac{\frac{dY}{dt}}{Y(1-Y)}\right]$ vs. $\ln\frac{1-Y}{Y}$ from Eq.7 would provide values of both 'n' and 'K' from slope and intercept respectively.

Although, SID method is able to provide both 'n' and 'Q' from a single dilatometric experiment, the technique requires a special programme to run dilatometer at a preset constant rate of shrinkage (CRS) mode. Several researchers[19-22] have replaced CRS with a special heating schedule to carry out SID experiments. However, the results obtained from SID technique were found to be inconsistent. The common practice in the isothermal sintering study is to correct the initial shrinkage data. Till date, no such effort has been made to incorporate shrinkage data correction in the SID method. This may have caused inconsistency in the end result. SID technique may also be useful to determine activation energy of sintering by using modified Dorn method[23], which is useful technique to calculate the activation energy of sintering [11].

From the above discussion, it can be found out that there is enormous scope to work in the field of sintering of ceramics. There is also a requirement to carry out a comparative study of the different models available and to find out their viability. At the same time, improvement in the SID data analysis is required to make this technique universally accepted. In the present investigation, the sintering kinetic study was carried out for 2Y-TZP/Al_2O_3 composite (2mol% yttria doped tetragonal zirconia added with 2 wt%Al_2O_3) using different available kinetic models and the results were compared to find out the usefulness of those models.

EXPERIMENTAL

Nano crystalline 2Y-TZP/Al_2O_3 (2mol% yttria doped tetragonal zirconia added with 2 wt%Al_2O_3) composite powder was used in the present investigation. Composite powder was synthesized following a method described elsewhere[24, 25]. In brief, mixed solution of zirconium oxy-chloride, yttrium and aluminium nitrate solution was co-precipitated at pH~ 9.0 in presences of 7.5 wt% polyethylene glycol and 10 wt% ammonium sulphate. Precipitate was washed with ammoniacal water and subsequently with alcohol. Dried precipitate was calcined at 700°C. Crystallite size and specific surface area of the calcined powder were 8 nm and 58 m^2/gm respectively. The calcined powder compacted isostatically at 200 MPa to a cylindrical shape (dia ~6mm, length 8 mm and green density ~ 43%TD (theoretical density)) was used for recording linear shrinkage in air during heating with a constant rate (400°C/h). The experiment was carried out in a double push rod dilatometer (TD5000SMAC Science, Japan) up to 1300°C. The approach for the data analysis was based on Johnson model[1] treating grain boundary diffusion (GBD) and volume diffusion (VD) separately. This assumption is valid only at the initial stage of sintering (5% shrinkage). For CRH experiment (T=ct where "c" is the heating rate), the general form of modified Johnson equation can be written as (Eq.8)-

$$y^{(\frac{1}{n}-1)}T\frac{dy}{dT}=A\exp(-\frac{Q}{RT}) \quad (8)$$

After integration and differentiation of Eq.8, Young and Cutler[2] made the following equation (Eq.9):

$$T\frac{dy}{dT}\approx A\exp(-\frac{nQ}{RT}) \quad (9)$$

Based on this equation, they concluded that a plot of $\ln\left[T\frac{dy}{dT}\right]$ vs. $\left(\frac{1}{T}\right)$ would give a straight line and slope of that straight line would be $\left(-\frac{nQ}{R}\right)$. Woolfrey and Bannister [11] proposed to plot $\left[T^2\frac{dy}{dT}\right]$ vs. 'y' to obtain a slope of $\left(\frac{nQ}{R}\right)$. Both these relations would be used in the present investigation for analysing shrinkage data. Relation given by Perez-Maqueda et al[5] would also be utilized for analysis of the sintering data.

For carrying out SID experiment, a special heating schedule was employed, in which the sample was held at temperatures 850, 890, 930, 970 and 1010°C for 30 minutes each. The ramping rate between the isothermal holdings was 10°C/min.

In the modified Dorn method, the instantaneous effect on shrinkage rate of a small step change in temperature is determined. If $\left[\frac{dy}{dT}\right]_{T_1}$ is the shrinkage rate at a temperature T_1 (K) just before the temperature change and $\left[\frac{dy}{dT}\right]_{T_2}$ is the rate of shrinkage at temperature T_2 just after the change, the activation energy of the process responsible for sintering is given by (Eq.10):

$$Q = \frac{RT_1T_2}{T_1 - T_2}\ln\left[\frac{\left(\frac{dy}{dt}\right)_{T_1}}{\left(\frac{dy}{dt}\right)_{T_2}}\right] \quad (10)$$

Where, R is the gas constant. In the present investigation, shrinkage rate obtained at end point of each isothermal step and at the beginning of next isothermal step was considered for determining the activation energy by Dorn method.

RESULTS:

CRH shrinkage analysis:

Shrinkage of the powder compact as a function of temperature during CRH experiment is shown in Fig.1. This shows that the shrinkage in the compact initiated at ~800°C (Point A in Fig.1). Sintering kinetics for the composite material has been analyzed in the initial sintering stage, i.e. within

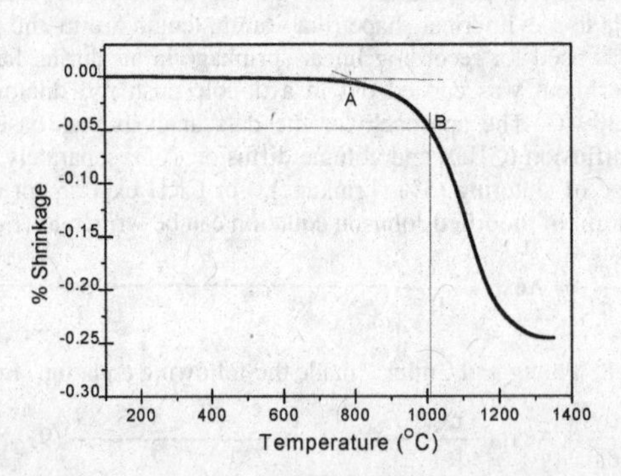

Fig.1 Shrinkage of compact of 2Y-TZP/Al_2O_3 composite powder as a function of temperature

the limit of initial 5% shrinkage (marked as point B). Figs.2 and 3 show the plots of $\ln\left[T\frac{dy}{dT}\right]$ vs. $\left(\frac{1}{T}\right)$ and $\left[T^2\frac{dy}{dT}\right]$ vs. 'y' respectively. It was observed that the plots were not completely linear over the whole range of densification. The situation may be viewed as two straight lines separated by a break point. The temperature corresponding to this break point is found to be ~900°C from both the figures. The average values of 'nQ' measured from the slope of the linear portion of the plots are indicated in the figures. In the lower temperature range (<900°C), 'nQ' values obtained from Young and Cutler model[2] (Fig.2) and Bannister model[11] (Fig.3) are found to be ~105 and 100 kJ/mol respectively. In the higher temperature zone (>900°C), the values of 'nQ' obtained from Fig.2 and 3 were 155 and 150 kJ/mol respectively. This indicates that the 'nQ' values obtained from two different models were in close proximity. However, from a single CRH experiment, values of 'n' and 'Q' can not be calculated separately. Prez-Maqueda et al[5] claimed that both these values can be calculated separately from a single CRH experiment. Following their equation (Eq.2), we have plotted $\frac{d(\ln\frac{dy}{dT})}{d(\ln y)}+\frac{d(\ln T)}{d(\ln y)}$ (considering a=-1) as a function of $\frac{d(\frac{1}{T})}{d(\ln y)}$ in Fig.4. As expected, this graph did not show any linear relation. Hence, neither 'Q' nor 'n' can be determined from this plot.

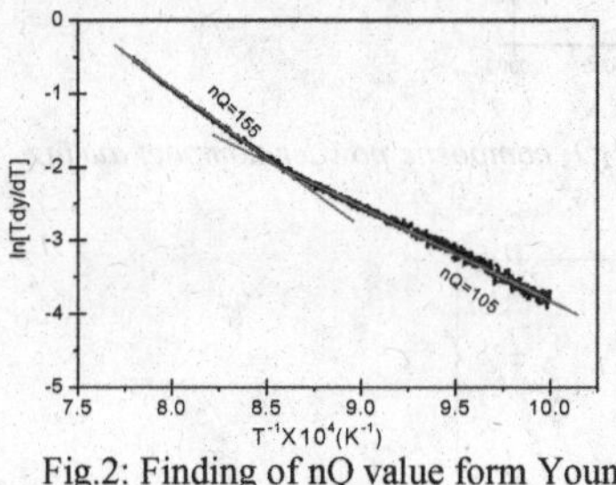

Fig.2: Finding of nQ value form Young and Cutler model[2]

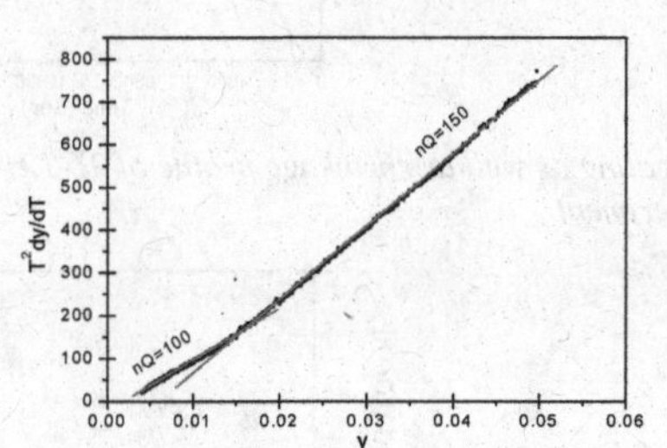

Fig.3: Finding of nQ value form Woolfrey and Bannister model[11]

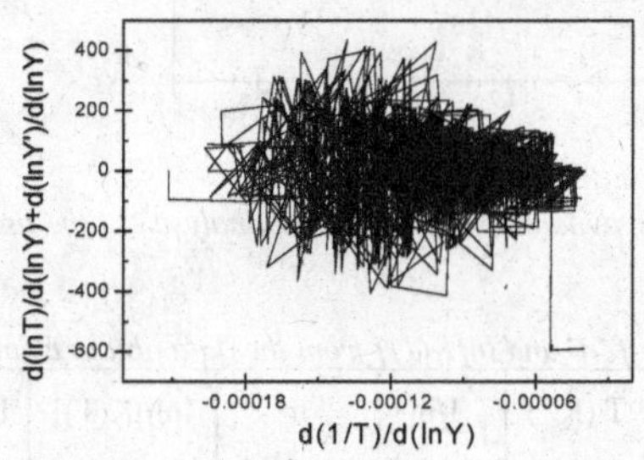

Fig.4: Prez-Maquedia plot for determining 'Q' and 'n' simultaneously.

SID method:

Shrinkage of the compact as a function of time and temperature during SID experiment is shown in Fig.5. Considering the shrinkage in the sample to be isotropic in nature, i.e.

$Y=\frac{L_O^3 - L_t^3}{L_O^3 - L_f^3}$, $\ln\left[\frac{\frac{dY}{dt}}{Y(1-Y)}\right]$ as a function of $\ln[\frac{1-Y}{Y}]$ (from Eq.7) is plotted (as shown in Fig.6) for the different isotheramal regions. A linear relationship has been observed for every isothermal region. Since, the shrinkage value in the 1010°C isotherm was exceeding 5% limit, data for this temperature was not considered for further discussion. The slope and intercept of the straight lines give '1/n' and 'ln[nK(T)]' respectively. From these, the values of 'n' and 'ln[K(T)]' have been calculated and listed in Table-1. Based on Eq.5, an Arrhenius plot of ln[K(T)] vs 1/T is drawn in Fig.7. From the slope of this linear plot, the activation energy was calculated and it was found to be 963±25 kJ/mol.

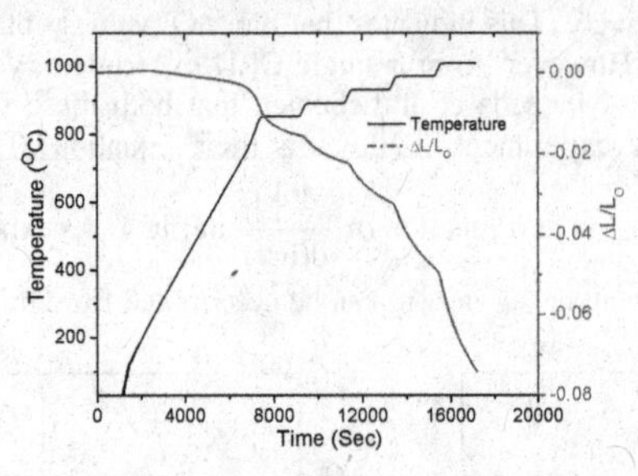

Fig.5 Heating as well as shrinkage profile of 2Y-TZP/Al_2O_3 composite powder compact during SID experiment

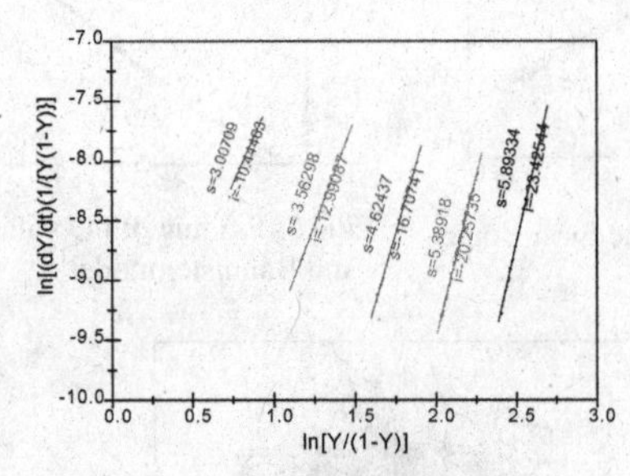

Fig.6: Sorrensen-Ming relation for dynamic shrinkage data analysis. In the figure, 's' and 'i' represent slope and intercept respectively.

Table-1 Calculation of 'n' and ln[K(T)] from the data obtain from Fig.6.

Temperature (°C)	1000/T (K^{-1})	1/n	n	ln[nK(T)]	ln[K(T)]
853	0.888	5.89	0.17	-23.42	-21.65
891	0.859	5.39	0.19	-20.25	-18.59
931	0.831	4.62	0.22	-16.70	-15.18
971	0.804	3.56	0.28	-12.99	-11.71

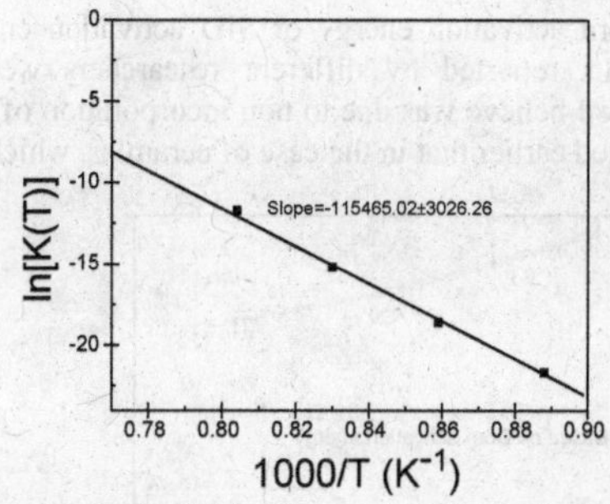

Fig.7: Arrhenius plot. Ln[K(T)] has been plotted as a function of inverse temperature.

Dorn method:

The value of activation energy of sintering in the different temperature zone obtained from Eq.10 is shown in Table-2. Value of activation energy of sintering is found to be marginally less at lower temperature. Surprisingly, the activation energy value obtained from the Dorn method was found to be almost half of the value measured from SID technique.

Table-2: Activation energy measured by Dorn method at different temperature interval

Temperature range (°C)		Activation energy (kJ/mol)
Starting temperature	End temperature	
853	891	440
891	931	475
931	971	505
971	1010	505

DISCUSSION:

Modified Johnson equation (Eq.8) has never been tried to calculate 'n' or 'Q' values, though it would provide more accurate result as compared to the value obtained from Young and Cutler model[2] or Woolfrey and Bannister equation[11]. In the present investigation, the value of activation energy obtained from both Dorn and SID techniques were used in the modified Johnson equation to calculate the 'n' values. For this, Eq.8 has been modified to the following form:

$$\ln[T\frac{dy}{dT}] + \frac{Q}{RT} = \ln A + \left(1 - \frac{1}{n}\right)\ln[y] \quad \text{------------------------ (11)}$$

Left hand side of the Eq.11 was plotted against ln[y] in Fig.8. From the slope of the straight line, the value of 'n' can be found out. Values of 'n' calculated from different equations have been listed in Table-3. This clearly indicates that the 'n' values obtained from SID technique, from modified

Johnson equation using either Dorn activation energy or SID activation energy were different. As mentioned earlier, the SID results reported by different researchers were found to be highly scattered[19-22]. This inconsistency we believe was due to non incorporation of correction for the initial shrinkage data. It has been mentioned earlier that in the case of ceramics, which have poor thermal

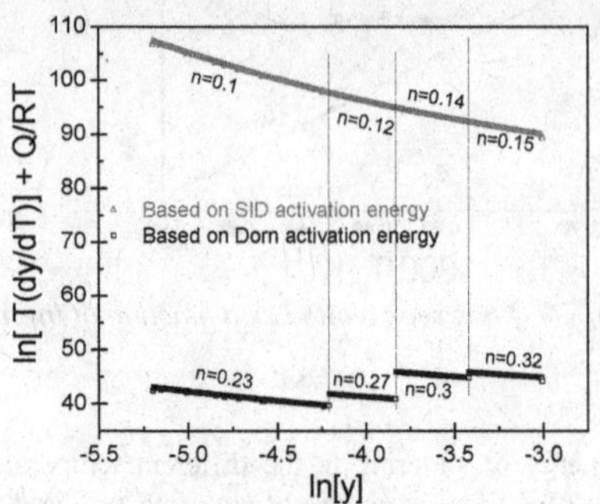

Fig.8. Graphical presentation of modified Johnson equation (Eq.11) for calculating 'n' values using activation energy obtained from SID and Dorn method.

Table-3 Value of 'n' measured from different equations using different 'Q' values.

Temperature range (°C)	'n' value from different techniques/models				
	From Young and Cultler model		From modified Johnson equation		From SID technique
	Using Activation energy from Dorn method	Using Activation energy from SID technique	Using Activation energy from Dorn method	Using Activation energy from SID technique	
Up to 891	0.23	0.11	0.22	0.1	0.17-0.19
891-931	0.30	0.15	0.25	0.12	0.19-0.22
931-971	0.31	0.16	0.29	0.14	0.22-0.28
971-1010	0.33	0.17	0.30	0.15	0.28 (971°C)

conductivity, the shrinkage initiates before reaching a steady state temperature. To overcome the shrinkage error caused by this phenomenon, several researchers suggested for initial shrinkage correction[7, 8, 13]. In the present work, following the technique adopted by Bacmann and Cizeron [13], initial shrinkage correction was made. Fig.6 is replotted incorporating this correction as shown in Fig.9. From the slope and intercept, values of 'n' and ln[K(T)] were measured. The 'n' values obtained from the SID technique were found to be almost similar to the values obtained from Young-Cutler model or modified Johnson method using Dorn activation energy. These values are plotted as a function of 1/T in Fig.10. 'n' vs 1/T plot clearly indicates a change in the sintering mechanism above 930°C. The average activation energy for sintering was found to be 420 (±30) and 550(±50) kJ/mol within 850°-890° and 850°-930°C temperature range respectively. These values were in close proximity to the values measured by Dorn method. Since, it has been observed in the present investigation that there was a change in sintering mechanism beyond 930°C, the activation energy for sintering in that region was not calculated from SID technique.

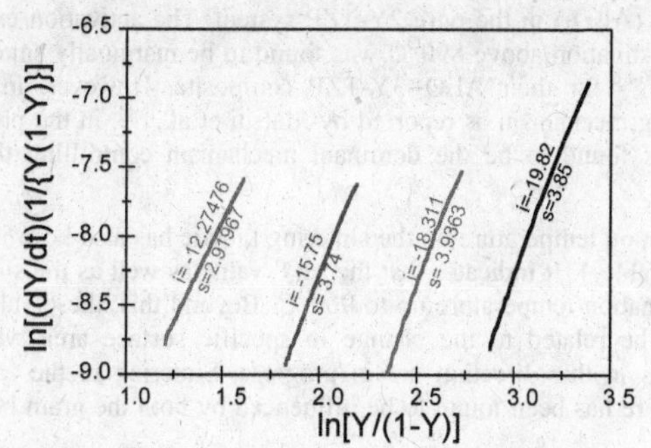

Fig.9: Replotting of Fig.6 incorporating the initial shrinkage correction.

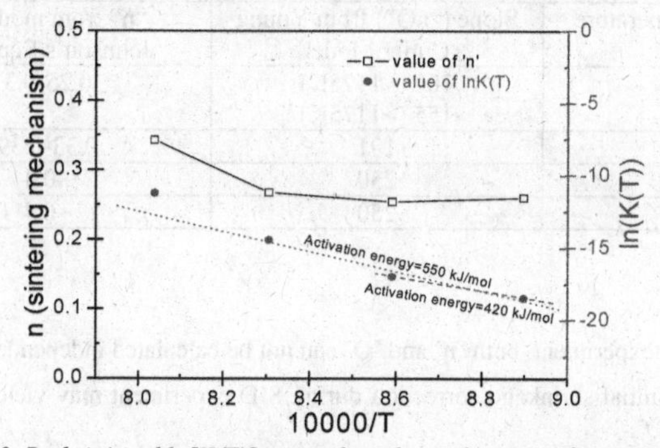

Fig.10. Both 'n' and ln[K(T)] were plotted as a function of inverse temperature.

The results obtained so far indicates that both the mechanism of sintering as well as the activation energy for sintering were changing as a function of sintering temperature. This manifested in the form of two different slopes in the Arhenius plot (Fig.2). During the initial stage of sintering, the value of 'n' was ~0.25. Value of n less than 0.33 indicates that other diffusion mechanism (apart from grain boundary diffusion) might be present at that stage of sintering. Rhodes[26] showed a grain boundary or surface diffusion controlled sintering mechanism at the initial stage of densification for an agglomerate free nanocrystalline yttria stabilized zirconia powder. Duran et al.[27, 28] postulated that the initial stage of sintering in the nano crystalline TZP was controlled by the combined surface and grain boundary diffusion mechanisms. Both Duran et al.[27, 28] and Theunissen et al.[29] observed non-linear Arrhenius plot in their CRH experiment. However, in their experiment, they crossed the 5% shrinkage limit. Hence, presence of different diffusion mechanism causing a change in slope was expected in their experiment. Simultaneous presence of more than one diffusion mechanisms during the sintering stage was also reported in recent times[30]. In the present investigation, however, a change in sintering mode is observed at the initial stage itself. Here, most likely, surface diffusion assisted grain boundary diffusion was the dominant controlling mechanism at the first part of initial stage of sintering. During the latter part, most likely, sintering have occurred by grain boundary diffusion. The calculated value of activation energy for sintering was found to be higher (>400 KJ/mol) as compared to 350 KJ/mol or less reported for TZP system in the literature [27-29]. This difference may have arisen because of the presence of a second phase (Al_2O_3) in the pure 2Y-TZP system. The activation energy for sintering obtained in the present investigation above 890°C was found to be marginally smaller than the value reported by Matsui et al.,[6-8] for their Al_2O_3-3Y-TZP composite. However, in place of volume diffusion controlled sintering mechanism as reported by Matsui et al.,[6-8], in the present investigation grain boundary diffusion is found to be the dominant mechanism controlling the initial stage of sintering.

The effect of calcination temperature on the sintering kinetic has also been investigated. Those results are summarized in Table-4. It indicates that the 'nQ' value as well as the sintering mechanism was dependent on the calcination temperature up to 900°C. Beyond this, these values remain almost constant. This effect may be related to the change in specific surface area with the calcination temperature. Further studies in this direction are in progress. Sintering in the case of the powder calcined at higher temperature has been found to be influenced by both the grain boundary as well as the volume diffusion.

Table 4: Effect of calcinations temperature on 'nQ' and 'n'.

Calcination Temperature	Slope ("nQ") from Young Cutler Model	"n" from modified Johnson's Equation
700°C	100 (<1175K) 155 (>1175K)	0.25-0.3
800°C	191	0.33-0.39
900°C	250	~0.4
1000°C	250	~0.4

CONCLUSIONS:

1. From single CRH experiment, both 'n' and 'Q' can not be calculated independently.
2. Incorporation of initial shrinkage correction during SID experiment may yield reliable value of 'n' and 'Q'.
3. In 2Y-TZP/Al_2O_3 composite, a change in sintering mechanism from combined surface and grain boundary diffusion to pure grain boundary diffusion has been observed in low temperature calcined powder.
4. At the initial stage, calcinations temperature influences the mechanism of sintering.

REFERENCES:

1. D.L.Johnson, *New method of obtaining volume, grainboundary and surface diffusion coefficients from sintering data.* J.Appl.Phys, 1969. **40**(1): p. 192-200.
2. W.S.Young and I.B.Cutler, *Initial sintering with constant rate of heating.* J.Am.Ceram.Soc, 1970. **53**(12): p. 659-663.
3. J.Wang and R.Raj, *Estimation of the activation energies for boundary diffusion from rate controlled sintering of pure alumina, and alumina doped with zirconia or titania.* J.Am.Ceram.Soc, 1990. **73**(5): p. 1172-1175.
4. H.Su and D.L.Johnson, *Master sintering curve: a practical approach to sintering.* J.Am.ceram.Soc, 1996. **79**(12): p. 3211-3217.
5. L.A.Prez-Maqueda, J.M.Cariado, and C.Real, *Kinetics of initial stage of sintering from shrinkage data: simultaneous determination of activation energy and kinetic model from a single non-isothermal experiment.* J.Am.ceram.Soc., 2002. **85**(4): p. 763-768.
6. K.Matsui, N.Ohmichi, and M.Ohgai, *Sintering kinetics at constant rate of heating.....* J.Am.ceram.Soc, 2005. **88**(12): p. 3346-3352.
7. K.Matsui, A.Matsumoto, M.Uehara, N.Enomoto, and J.Hojo, *Sintering kinetics at isothermal shrinkage: effect of specific surface area on the initial sintering stage of fine zirconia powder.* J.Am.Ceram.Soc., 2007. **90**(1): p. 44-49.
8. K.Matsui, K.Tanaka, T.Yamakawa, M.Uehara, N.Enomoto, and J.Hojo, *Sintering kinetics at isothermal shrinkage: II, effect of Y_2O_3 concentration on the initial sintering stage of fine zirconia powder.* J.Am.Ceram.Soc, 2007. **90**(2): p. 443-447.
9. P.Tanev, S.Koruderlieva, C.Leach, and B.Russeva, *Kinetics od sintering of α–alumina oxide derived from alumina-ammonium alum.* J.Mater.Sci.Lett, 1995. **14**: p. 668-669.
10. W.Young, S.Rasmussen, and I.B.Cutler. *Ultrafine grained ceramics* in *15th Sagamore Army Materials Research Conference*. 1970. Syracuse, N.Y: Syracuse University Press.
11. L. Woolfrey and M.J.Bannister, *Non-isothermal techniques for studying initial stage sintering.* J.Am.Ceram.Soc, 1972. **55**: p. 390-394.
12. D.Lahiri, S.V.R. Rao, G.V.S.H. Rao, and R.K.Srivastava, *Study on sintering kinetics and activation energy of UO_2 pellets using three different methods.* J.Nucl.Mat., 2006. **357**: p. 88-96.
13. J.J.Bacmann and G.Cizeron, *Dorn method in the study of initial phase of uranium dioxide sintering.* J.Am.Ceram.Soc, 1968. **51**(4): p. 209-12.
14. R.M. Laine, J.C. Marchal, H.P. Sun, and X.Q. Pan, *Nano-α-Al2O3 by liquid-feed flame spray pyrolysis.* Nature materials, 2006. **5**: p. 710-712.
15. O.T.Sorensen, *Interpretation of quasi-isothermal thermogravimetric weight curves.* Thermochim Acta, 1979. **29**: p. 211-14.
16. O.T.Sorensen, *Thermogravimetric and dilatometric studies using stepwise isothermal analysis and related techniques.* J.Thermal Analysis, 1992. **38**(1-2): p. 213-228.

17. M.E.S.Ali and O.T.Sorensen and S. Meriani, *Dilatometric sintering studies on ceria-zirconia powders.* Science of ceramics. 1984. **12**: p. 355-360.

18. G.Y.Meng and O.T.Sorensen, *Kinetics analysis on low temperature sinter process for Y-TZP ceramics.* Advanced structural materials, ed. Y. Han. Vol. 2. 1991, Amsterdam: Elsevier Science Publisher. 369-374.

19. H.-t. Wang, X.-q. Liu, F.-l. Chen, G.-y. Meng, and O.T.Sorensen, *Kinetics and mechanism of a sintering process for macroporous alumina ceramics by extrusion.* J.Am.Ceram.Soc, 1998. **81**(3): p. 781-784.

20. Y.-f. Liu, X.-q. Liu, S-w.Tao, G-y.Meng, and O.T.Sorensen, *Kinetics of the reactive sintering of kaolinite-aluminium hydroxide extrudate.* Ceramic International, 2002. **28**: p. 479-486.

21. X.Zhang, X-q.Liu, and G-y.Meng, *Sintering kinetics of porous ceramics from natural dolomite.* J.Am.Ceram.Soc, 2005. **88**(7): p. 1826-1830.

22. R.Yan, F.Chu, Q.Ma, X-q.Liu, and G-y.Meng, *Sintering kinetics of samarium doped ceria with addition of cobalt oxide.* Materials Letters, 2006. **60**: p. 3605-3609.

23. J.J.Bacmann and G.Cizeron, *Dorn method in the study of initial phase of uranium dioxide sintering.* J.Am.Ceram.Soc, 1968. **51**(4): p. 209-212.

24. A.Ghosh, A.K.Suri, B.T.Rao, and T.R.Ramamohan, *Low temperature sintering and mechanical property evaluation of nanocrystalline 8mol% yttria fully stabilized zirconia.* . J.Am.Ceram.Soc, 2007. **90**(7): p. 2015-2023.

25. A.Ghosh, A.K.Suri, B.T.Rao, and T.R.Ramamohan, *Synthesis of nanocrystalline sinteractive 3Y-TZP powder in presence of ammonium sulphate and poly ethylene glycol* Adv.Appl.Ceram.: p. communicated.

26. W.H.Rhodes, *Agglomerate and particle size effect on sintering yttria stabilized zirconia* J.Am.Ceram.Soc., 1981. **64**(1): p. 19-22.

27. P. Duran, P.Reico, J.R.Jurado, C.Pascaul, and C.Moure, *Preparation sintering and properties of translucent Er2O3- dopeed tetragonal zirconia.* J.Am.Ceram.Soc, 1989. **72**(11): p. 2088-93.

28. P.Duran, M.Villegas, F.Capel, P.Recio, and C.Moure, *Low temperature sintering and microstructural development of nanocrystalline Y-TZP powders.* J.Eur.Ceram.Soc., 1996. **16**: p. 945-52.

29. G.S.A.M.Theunissen, A.J.A.Winnubst, and A.J.Burggraaf, *Sintering kinetics and microstructure development of nano scale Y-TZP ceramics.* J.Eur.Ceram.Soc., 1993. **11**: p. 315-324.

30. G.Suarez, L.B.Garrido, and E.F.Aglietti, *Sintering kinetics of 8Y-cubic zirconia: cation diffusion coefficient.* Materials chemistry and physics, 2008. **110**: p. 370-375.

INVESTIGATIONS OF PHENOLIC RESINS AS CARBON PRECURSORS FOR C-FIBER REINFORCED COMPOSITES

H. Mucha*, B. Wielage**, W. Krenkel*

* Department of Ceramic Materials Engineering, University of Bayreuth, 95440 Bayreuth, Germany
** Chair of Composite Materials and Surface Technology, Chemnitz University of Technology, 09107 Chemnitz, Germany

ABSTRACT

Different Carbon matrix precursors are used for the fabrication of C/C-SiC materials by the liquid silicon infiltration (LSI) route. Phenolic resins represent an important class of suited resins which includes the resol and novolac types. For these resins a universal chemical and structural point of view is adopted. It is based on the aromatic bonding state and leads to the concepts for the carbon yield and the basic structural units after carbonisation. The mechanisms that allow substituents to control the reactivity of the aromatic ring structures, control the phenol-formaldehyde-addition as well, which predetermines the character of its basic resin product (resol or novolac). Direct curing (temperature or acid) is possible for resols. Novolacs are thermoplastics and not curable without additives. With suited addives, novolacs can be cured indirectly. Novolac with an additive (eg. hexamethylentetramine) form a product which is cured at elevated temperatures and hence forming characteristic bridges in case-typical ratios by condensation reactions and thereby releasing gaseous reaction products. The primary bridges show ranges of thermal stability and exhibit different mechanical strengths. This accounts also for the phase of thermal transformations when bridges transform into other bridge types or disintegrate. These thermally induced transformations are investigated for a few resol type resins on a macroscopic mechanical level using the Dynamical Mechanical Thermal Analysis (DMTA) and characterised supplementary by thermo-gravimetric means. The generated porosity is a result of the fabrication history and subjected to microstructural investigations. Furthermore attention is paid to the effect of porosity for the Si-infiltration.

INTRODUCTION

The availability of phenolic resins in a defined chemical state is required for the fabrication of CFRP, C/C- and C/C-SiC materials in a reproducible way. Nevertheless microscopic investigations of the obtained matrix show that it is not a homogeneous, dense component and it exhibits porosity of different shape and size. The occurrence of porosity is depending on the type of resin and its processing conditions. Pore sizes may range from macro- to nanoscale and can be revealed by suited preparation and analysis technologies. Reasons for the generation of pores originate in the resin`s molecular constitution, which is essential for the initial curing and the subsequent thermal conversions of the bridges and the cracking processes at higher temperatures. The gaseous products from the reaction are released from or embedded into the resin depending on the resin`s constitution. To get accustomed to the class of resins it is required to know the underlying principles of bonding.

Phenolic resins are multi particle systems. Central field models that are suited to describe atoms are not adequate to describe the electron distributions within molecules and multi particle formalisms have to be applied. Unfortunately such multi particle systems can not be solved precisely and simplified model approaches are required. A multitude of approaches has been developed in the past and it is necessary to get an overview about the required assumptions and their conditions of validity. The presently most comprehensive, modern representation of the chemistry of phenols is edited by Z. Rappoport[1] in 2003. It covers amongst others the Hartree-Fock, Post Hartree-Fock and Density Functional Theory approaches. Calculated and experimentally obtained properties of phenols are electron density distributions, lengths of bonds and angles between bonds. By these means microscopic parameters and macroscopic properties are correlated. Some aspects of the behaviour of phenols have been investigated on that microscopic base and are visualised for a better understanding. The electron density distribution of the ring structure is visualised for different substituents and for cases where a phenol forms a complex with an akali atom (catalyst for Aldol reaction) or phenol interacts with water (solvent for many phenolic resins). Hydrogen bridges are temperature and also polarity sensitive and induce charge exchanges with the electronic system of the ring; this coupling controls the reactivity of the phenolic system and therefore the thermal stability of the bridges and also their transformations. The reaction conditions of the phenol-formaldehyde addition attribute a permanent character to the resin prepolymers.

Theoretical descriptions of the resins assume a homogeneous material distribution, but porosity is occuring. Porosity is caused by kinetic limitations and are mostly accompanied to a release of solvents, reaction products from curing or to crack products of aliphatic compounds. A series of reports from Cypres and Bettens[2] considers a recondensation of nonaromatic components into aromatic components at carbonisation temperatures.

The concept of aromatic bonding is of outstanding importance for the description of the phases of curing, the phase of bridge transformation at higher and the phase of disintegration at more elevated temperatures. It is also a key issue to understand the carbonisation processes. The delocalisation of the electron within the bond is a quantum mechanical effect and attributes an extremely lowered energy to the delocalised, ie. the aromatic case. As a result aromatic structures do not disintegrate even at elevated (carbonisation) temperatures (inert atmosphere), whereas the aliphatic and alicyclic compounds do and finally serve no carbon yield. The carbon materials can be classified according to the general system of carbon materials by the concept of basic structural units (BSU) and local molecular order (LMO), terms coined by A. Oberlin[3]. Depending on the local arrangement of these BSUs well ordered, extended structures (large LMO) or crumbled sheet like, unorganised structures (small LMO) or any structure in between may be realised. Phenolic resins as other aromatic materials with a 3-dimensional network structure result in amorphous (=small LMO) carbon materials. In case of meso-phase pitch based materials (liquid crystalline materials) a high local molecular order is found after carbonisation. The presence of a large amount of network promoting elements (=strong network) is a reason for obtaining a badly ordered carbonaceous material. This usually can not be changed later even under high temperature treatments (not graphitising materials). The graphitising process usually has only a small time and temperature window to start. Many large LMO materials fall into such window and therefore can be graphitized under high temperature treatments[4]. The mentioned structural effects are not easily detectable as high resolution lattice imaging of the LMO is necessary, which means that high resolution transmission electron microscopy (HRTEM) is required.

The classical TEM preparation technology does not allow fast and pinpointed thin film investigations as they are needed for systematic studies of the porosity and of the local molecular order. For this reason the focused ion beam (FIB) technology is applied here and adapted to the preparation of C-fibre reinforced composites. Furthermore the FIB-TEM preparation technique reduces mechanically induced artefacts.

The thermally induced changes induced to the bridge structure on the molecular level should influence the macroscopic mechanical properties of that resin. For matrix dominated mechanical properties 3-point bending geometry is most suited. Therefore the the Dynamical Mechanical Thermal Analysis is used as a tool to reveal the temperature dependence of the elastic modulus.

EXPERIMENTAL

Phenolic resins are used as matrices for C-fiber reinforced composites which are in house fabricated and complemented by commercial composites. The in house fabrication of composites also applies modified versions of the starting resin. Three different types of commercial phenolic resins are subjected to produce CFRPs via an autoclave route. A part of this CFRP material is kept for later reference and testing whereas about 2/3 is subjected to a pyrolysis treatment in order to create C/C state material. One half of the C/C material is kept for reference and subsequent testing whereas the other half undergoes a liquid silicon infiltration (LSI-route) to form C/C-SiC composites. Therefore the 3 processing states derive materials from exactly the same starting resin and even the same CFRP material. This procedure is necessary to achieve comparability because the as received commercial phenolic resins are a low cost products and usually characterised mainly by technological properties and not on a scientific base.

Resin modifications aim to influence relevant resin parameters. The type and the amount of solvents are varied (solvent substituents) first and approaches to change the network design by a catalytic modification of the resin are performed.

The microstructural characterisation (optical microscope, SEM, TEM, FIB) of the composites covers the matrix and fibre/matrix interface areas and monitors the range from macroscopical dimensions down to the nano scale. Special attention is paid to the matrix porosity, their local distribution and to the occurrence of indications for stress graphitisation in C/C and C/C-SiC materials. The microstructural analysis is improved by applying the Focused Ion Beam technology as a reliable imaging method and deriving a powerful TEM-thin film preparation technique from it to improve the resolution by reducing the film thickness and to open matrix porosity on one or even on two sides to visualise them directly in the sub micrometer range.

The thermal investigations consist in a thermogravimetric characterisation (Netzsch TG 409) of the available resins for optimising the time–temperature profiles and DMTA (GABO EPLEXOR 150N, 1000°C) studies. In the DMTA experiments the CFRP samples (35mm x 7 mm x 3 mm) are subjected to in-situ 3 point bending tests under an inert gas atmosphere (ambient pressure) in a temperature controlled furnace, which is a part of the DMTA equipment and following a pyrolysis temperature profile up to 1000 °C.

THERMOGRAVIMETRIC RESIN ANALYSES

Three different resins (PV 1420, E97783, E 95308, all made by Dynea GmbH, Germany) were investigated by TG in the as received condition (Figure 1). After the removal of water (E97783 dest.) the mass loss decreased but the viscosity increased more than acceptable.

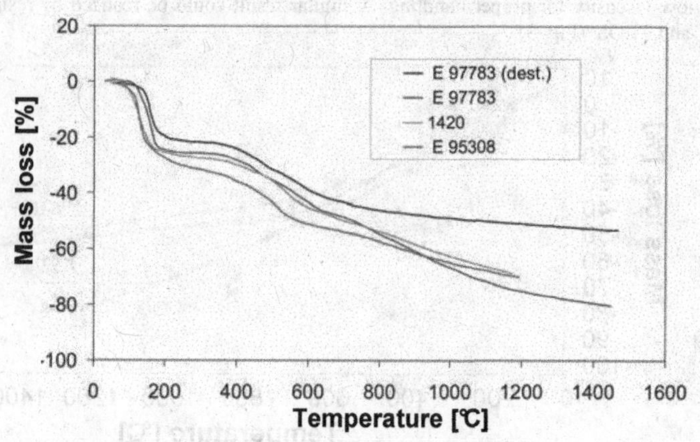

Figure 1. Temperature dependence of mass losses for as received and water free resins

The resin of type E97783 appeared as most promising and was subjected to further modifications. These modifications include a) the removal of the solvent water (E97783 dest.), b) solvent substitutions (exchange water to acetone, ethlenacetate, diethylether, tetrahydrofurane, furfurylalcohole, Figure 2) and c) realising modified network conditions by catalytic amounts of tetraetoxyorthotitanate - TEOS, tetraethoxyorthosilcate – TEOS, (Figure 3). The resin modifications had been performed on laboratory scale by the Chair of Inorganic Chemistry (Prof. H. Lang), Chemnitz University of Technology.

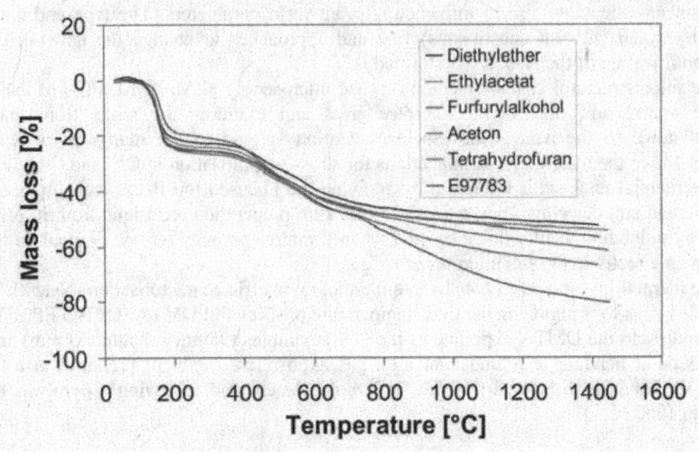

Figure 2. Temperature dependence of mass losses for as received and derived solvent substituted resins

The selected solvents reduced the mass losses compared to original water solved resin and retained a sufficient low viscosity for proper handling. A similar result could be realised by resin modifications via TEOT and TEOS (Fig. 3)

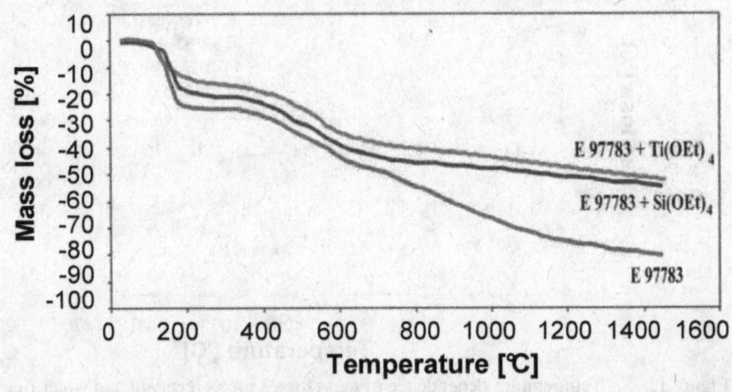

Figure 3. Temperature dependence of mass losses for as received and derived chemically modified Types (TEOT, TEOS)

SAMPLE FABRICATION

The sample fabrication (Figure 4) takes place by gravity infiltration of resin into a mould which is filled with a stack of precisely cut and aligned carbon fibre cloths (0°/90°). The infiltration is followed by a curing phase in an autoclave (50 bar, Argon, usually 160°C). Next, after removing the small CFRP-plate from the mould, the pyrolysis takes place in muffle type furnace at maximum temperature of 900 °C under ambient pressure and in Argon atmosphere. It is followed by a liquid silicon infiltration in a vacuum chamber with a graphite furnace at about 1550 °C. Similar pyrolysis experiments starting from already cured CFRP materials are also performed in the furnace of the DMTA machine, parallel to the mechanical measurements.

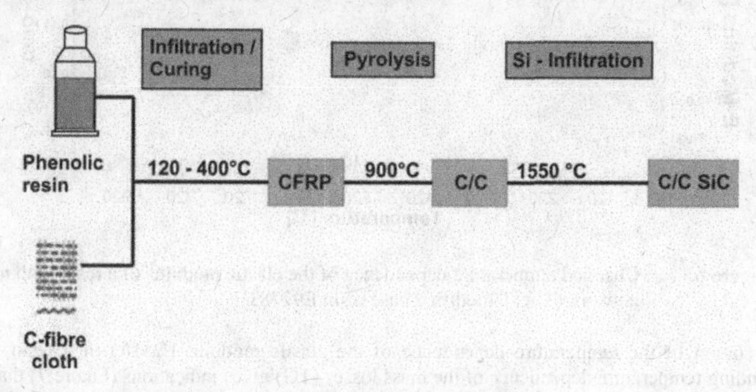

Figure 4. Sample fabrication steps, CFRP, C/C, C/C-SiC

DYNAMICAL MECHANICAL THERMAL ANALYSES (DMTA)

The DMTA technique and equipment allows the determination of the temperature dependence of the elastic modulus and mechanical damping (tanδ) of the test samples under load in 3-point bending geometry. As pyrolysis experiments can be performed in-situ in the DMTA furnace, any resultant thermally induced weakening or strengthening of the network can be detected on the macroscopic level and represented as a temperature dependence of the elastic modulus. Investigations of materials derived from the same base resin (E97783) by solvent substitution lead to results as depicted in Figure 5. Their temperature dependence is very similar and differs mainly in their amplitudes. In this case of resin vari-

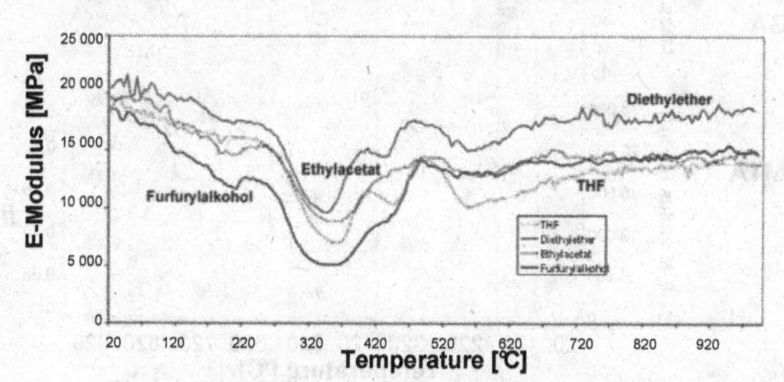

Figure 5. Similar temperature dependence of E- Modulus for solvent substituted resins

ations no significant differences among the obtained networks are expected (= same class) and the resultant temperature dependence of the elastic moduli do not differ significantly. Modified network conditions are expected in case of a resin modifications by TEOT or TEOS. As a result the network character changes and the temperature dependence of the elastic-modulus is reflecting this fact (Figure 6).

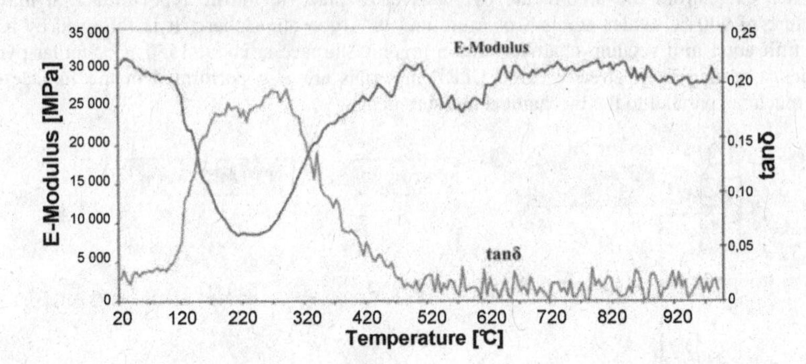

Figure 6. Changed temperature dependence of the elastic modulus of a resin with modified network (TEOT modified base resin E97783)

The correlation of the temperature dependence of the elastic modulus (DMA) of a resin with the corresponding temperature dependence of the mass losses (TG) gives indications (Figure 7) that 3 main reaction phases exist: 1.) a curing phase below 180 °C, 2.) a bridge/network transformation phase (180 °C - 400 °C) and a cracking phase above 400 °C. In the curing phase mass losses are essential and correlated with changes stiffness whereas in the transformation phase the stiffness changes are almost independent of mass losses. Above a certain temperature which is for the E97783-based resin in the range of 350 °C - 400 °C the transformations correspond to mass losses, i.e. the cracking phase starts.

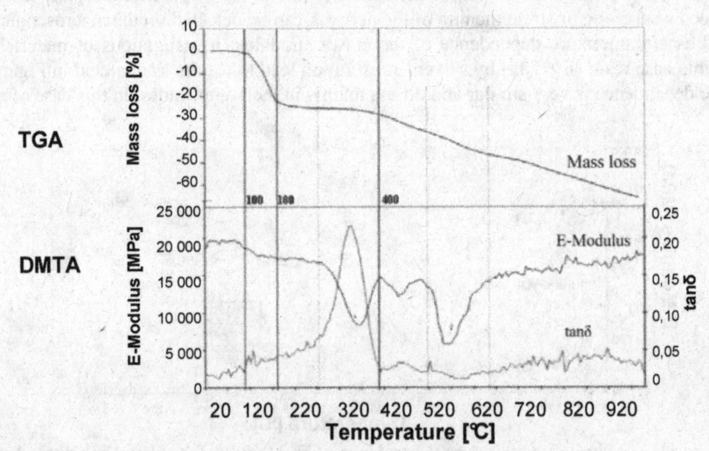

Figure 7. Thermal transformation ranges (E 97783)

MICROSTRUCTURE ANALYSES

Optical microscopy visualises two extreme examples of phenolic resin structures (resol type). Those of resols which form matrices in CFRPs that develop macroscopic porosity (PV1420, Figure 8a) and those (E95308, Figure 8b) which appear dense on the macroscopic scale and compensate the curing shrinkage in by matrix cracking within the rigid frame of the fibre reinforcement.

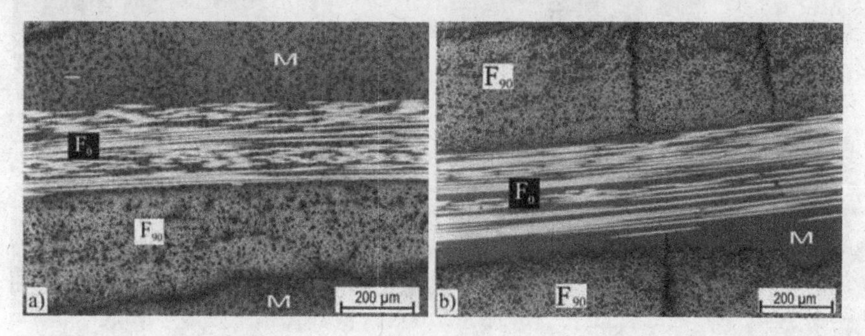

Figure 8. Optical micrographs of resol based CFRPs, a) PV1420 b) E95308
M = matrix, F_0, F_{90} = Fibre, resin maker: Dynea GmbH, Germany

For practical use a resol that is gaining an intermediate structure (Figure 9) in the CFRP state is most beneficial. It is realised for example by the resin E97783, which was already subject of other characterisations. It shows a narrow, open crack system (first segmentation), foamy zones near or inside cracks and closed porosity mainly on submicron scale in the matrix rich areas. A reliable preparation of such foamy zones is possible when they are located behind a fibre laying parallel to the surface as indicated in Figure 10. Continued polishing thins the remaining fibre and induces its final removeal which terminates the preparation. The highest magnification is applied at the foam region in Figure 10.

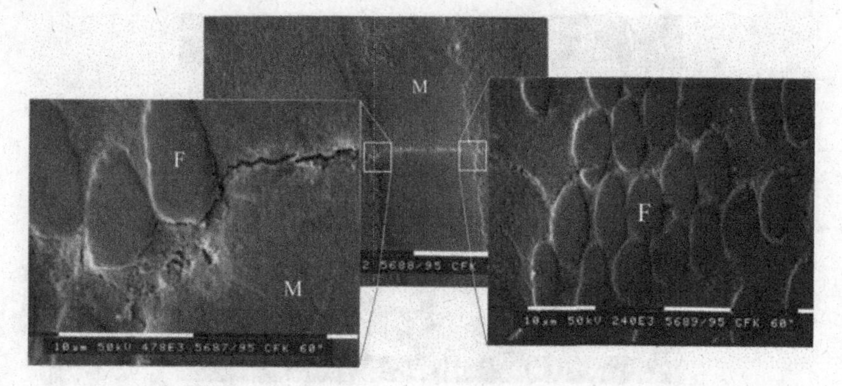

Figure 9. Optical micrograph of the structure of a resol based CFRP
M = Matrix, F = Fibre

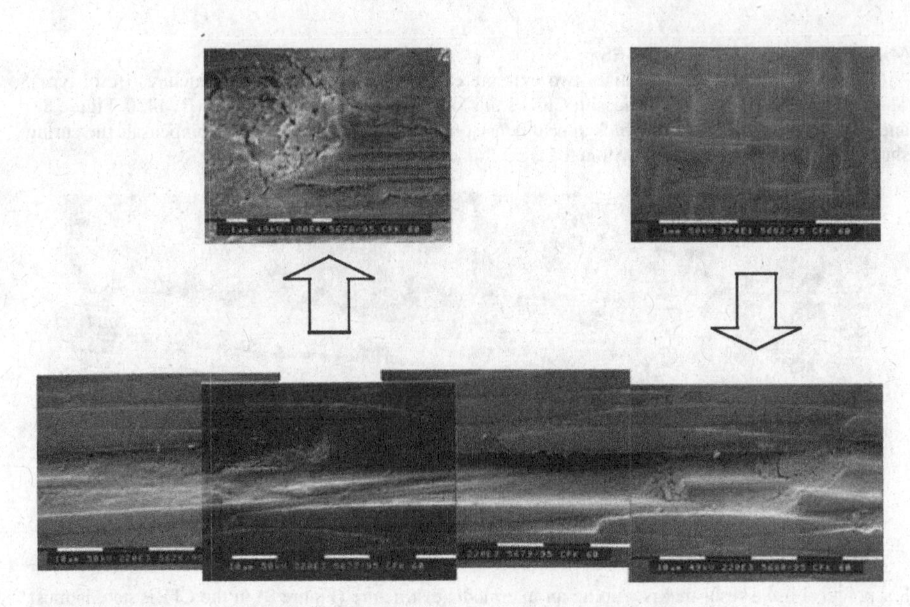

Figure 10. Preparation of foamy areas in resol based CFRP

The crack structure indicates the former function of this area as a drain for gaseous curing products, which were generated within the bulky matrix areas, collected and transferred to a connected open channel system (Figure 11). The porosity in bulk matrix areas which are adjacent to the foams and cracks in CFRPs (E97783 type resins) is reduced by several orders of magnitude compared to foam.

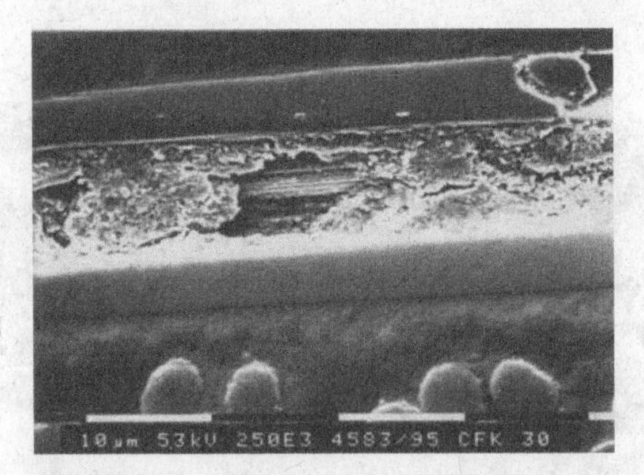

Figure 11. Connection of a foam area to an open channel system (CFRP)

The pyrolysis intensifies the generation of gaseous reaction products as well as the amount of open and closed porosity within the matrix as C/C composites. As a matter of fact already in the CFRP state existing porosity is not redeveloping during pyrolysis. The first segmentation cracks in the CFRP state become zones of reduced fibre density in the Carbon/Carbon (C/C) state and include open channel systems. Additionally a second segmentation is occurring in the C/C state (Figure 12).

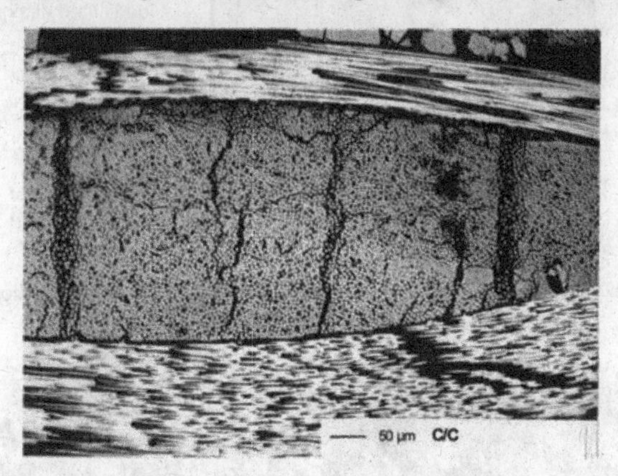

Figure 12. Zones of reduced fibre density and second segmentation in C/C

The first segmentation cracks take over the function of drains for pyrolysis gases and give indications for a partial recondensation. The average pore size, even in densely appearing matrix areas, is significantly higher than at the corresponding CFRP locations. The production of pyrolysis gas takes place all over the matrix volume, but significant differences of the porosity level are occurring. The local change of porosity in the transition area, from the minimum level of porosity to a foam area, is of great interest. The visualisation of such fine structures is possible by transmission electron microscopy and an efficient TEM thin film preparation method. For this purpose the FIB-TEM thin film lift-out preparation technique[5] was introduced and applied. Free standing composite films are produced from C/C lift-out samples in an arrangement as displayed in Figure 13. The Ga-Ion beam is directed almost perpendicular to the front face of the holder and used for pinpointed FIB-thinning. The thin films are prepared wit a Hitachi FB 2100 FIB machine equipped by a micromanipulator system. It succeeded to open nanopores within the preparation area on one or even on both sides (Figure 14.). The TEM micrograph (Figure 14a) shows an area adjacent to a foam area (upper right) and reduced porosity in the left part. The marked area is enlarged in Figure 14b, which depicts a "huge" pore (<100µm) which now is open on both sides and a large number of smaller pores. Due to the very thin FIB/TEM lamella the pores are almost never lying upon each other and can be identified separately on the TEM micrograph. It is to be assumed that these almost equally distributed pores (Figure 14 b) had been of closed type and accumulated locally generated pyrolysis gas.

Figure 15 resolves an area around a HTA 5131 fiber which is embedded in a C matrix derived from resin E97783. The pores adjacent to the fibre show a significant deformation from the spherical shape. This indicates the existence of significant stresses and strains near the fibre in the resin. By deformation of the pores the resin succeded to eliminate stress peaks and prevent local stress graphitisation which was expected to occur.

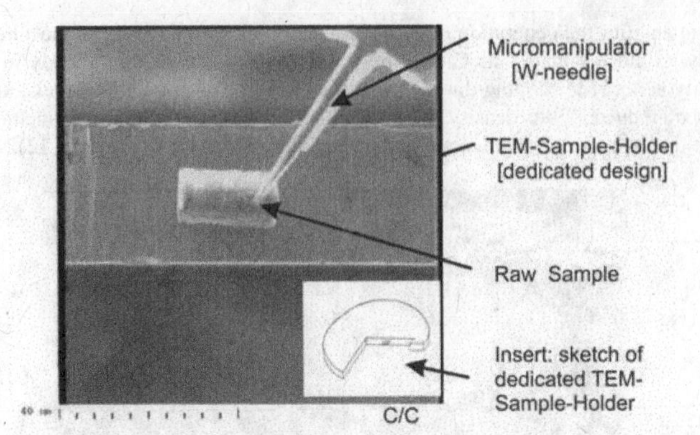

Figure 13. Arrangement of a free standing lift-out C/C sample on a dedicated FIB/TEM sample holder before final thinning /Muc05/

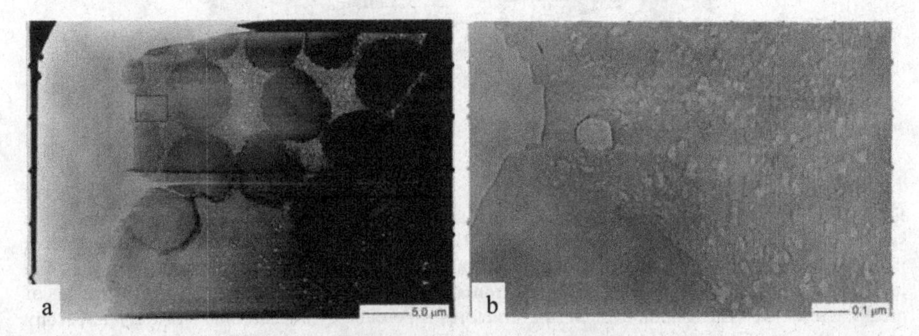

Figure 14. TEM micrograph of a C/C sample derived from E97783 resin, fibre: T800

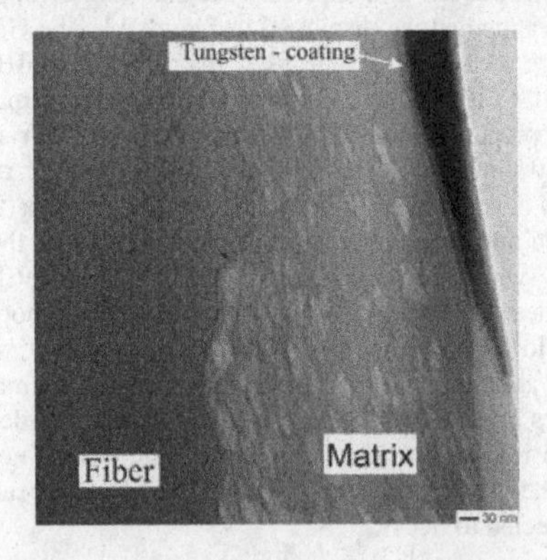

Figure 15. TEM micrograph of a C/C sample derived from E97783 resin, fibre HTA 5131

DISCUSSION AND CONCLUSIONS

The knowledge of the chemical constitution of phenolic resins on a microscopic base as well as that of the mechanisms of the carbonisation process are necessary to predict the performance of modified resins. It helps to systematize the multitude of possible chemical variations and their attributed properties. It justifies strategies to modify compounds with respect to their expected their properties but does not give precise guidelines.

The experimental approach aims to reveal correlations between microcopic mechanisms and macroscopic performance. The substitution of solvents and the network modifications (TEOS, TEOT) of the resins represent simple approaches to reveal and systemise the interdependences between microscopic and macroscopic parameters. In a first approximation the DMTA investigations confirm a correlation of the network type with the temperature characteristics of the elastic modulus.
The viscoelastic properties of uncured, cured and transforming resins and their temperature dependence originate from microscopic bonding effects even if they can not be attributed directly to single effects. The DMTA analyses are able to measure a macroscopic property (elastic modulus, damping) and correlate it to changes of its microscopic origin. The correlation of DMTA- with TG-measurements in the same temperature range serves additional information and affirms the postulated partitioning of the reactions in the 3 phases of curing (<180) / bridge transformations / (>400°)cracking.

The microstructural investigations of CFRPs based on different resins show a correlation between resin type and porosity. This indicates, that a significant influence of the curing reactions on the structure exists and that also a chance exists to control the microstructure by a suitable chemical resin design. The reaction gas transport in both CFRP and C/C materials is not uniform. They drain with locally intensive streams and foam zones but also show calm areas with relatively small closed pores. Specially these calm areas in C/C materials are investigated in more detail by TEM in order to find the smallest pore sizes. The applied FIB/TEM lamella preparation was able to produce such thin lamellae that many pores were opened on both sides and the TEM electron beam could create images of high contrasts, which could not be obtained by conventional TEM preparation. Pores with diameters of less than 20 nm diameter were detected and quite uniformly distributed in the analysis areas. Perhaps they can be considered as reservoirs filled with curing gas.

TEM investigations of matrix areas directly adjacent to Carbon fibres show submicron pores in these regions as well. But here they are strongly deformed and adopt lens like shapes. This indicates the existence of local stresses and strains in the matrix. Such conditions do not allow the accumulation of sufficiently high stresses to induce a stress graphitisation. The level of porosity in all these C/C materials based on E97783 resins is still too high in order to induce the stress graphitization.

The porosity takes over a key function in C/C materials. It is not intensively investigated and not efficiently controlled, presently. This is even more surprising as a tailoring of the open and closed porosity would allow to make use of C-foam as an easily accessible carbon source for the reaction with liquid silicon. This can reduce the Si-infiltration time and minimise the C-fiber damages due to reactions with silicon.

REFERENCES

[1] Phenols – The Chemistry of Phenols , 2 Bde; Hrsg.: Rappoport, Z., Wiley, 2003

[2] Cypres, R.; Bettens, B.: Mechanismes de fragmentation pyrolytique du Phenol et des Cresols, Tetrahedron, Vol. 30, pp. 1253-1260, 1974
Cypres, R.; Bettens, B.: Pyrolyse Thermique des [14C] et [3H] ortho et para-Cresols, Tetrahedron, Vol. 31, pp. 353-359, 1975
Cypres, R.; Bettens, B.: La formation de la plupart des composes aromatiques produits lors de la pyrolyse du phenol, ne fait pas intervenir le carbonne porteur de la fonction hydroxyle, Tetrahedron, Vol. 31, pp. 359-365, 1975

[3] Oberlin, A.: High-Resolution TEM Studies of Carbonization and Graphitization; in: Chemistry and

Physics of Carbon, Bd. 22, S. 1 - 144, Hrsg. Peter A. Thrower, Marcel Dekker Inc, New York, Basel, 1989 (TIB: ZN 1625,22)

4 Fitzer, E. ; Manocha, L. M.: Carbon Reinforcements and Carbon/Carbon Composites, Springer Verlag, Berlin,1998, ISBN 3-540-62933-5

5 Mucha, H.; Kato, T.; Arai, S.; Saka, H.; Kuroda, K.; Wielage, B.: Focused Ion Beam Preparation Techniques dedicated for the Fabrication of TEM-Lamellae of fibre-reinforced composites; Journal of Electron Microscopy, Bd. 54 (1), 2005, Heft 1, Seiten 1-7

ALUMINUM NITRIDE MULTI-WALLED NANOTUBE (MWNTs) NANOCOMPOSITE BY DIRECT IN-SITU GROWTH OF CNTs ON ALUMINUM NITRIDE PARTICLES

Amit Datye, Kuang-Hsi Wu, and S. Kulkarni
Mechanical and Materials Engineering
Florida International University
Miami, Florida 33174

H. T. Lin
Oak Ridge National Laboratory
Oak Ridge, TN 37831-6068

J. Vleugels
Department of Metallurgy and Materials Engineering
Katholieke Universiteit Leuven
Leuven, Belgium

Wenzhi Li, Latha Kumari
Department of Physics
Florida International University
Miami, Florida 33174

ABSTRACT

Aluminum Nitride (AlN) – Carbon Nanotube (CNT) composites were prepared to increase the fracture toughness, flexural strength and thermal conductivity of AlN based ceramics by direct in-situ growth of CNTs on the mixtures using Chemical Vapor Deposition (CVD) technique. The AlN- 3 wt% Yttria (Y_2O_3) base mixtures with in-situ grown CNTs were then spark plasma sintered at 1950°C in vacuum under a pressure of 60MPa. The Vickers Hardness decreased with the addition of CNTs in the matrix whereas the indentation fracture toughness increased with addition of CNTs. The flexural strength increased for mixtures with 7 wt% CNTs by comparison to base AlN-3 wt% Y_2O_3 samples.

Detailed electron microscopy analysis of the Aluminum Nitride-CNT powders shows uniform distribution of CNTs in the matrix, without the formation of agglomerates frequently seen with traditional ex-situ mixing of CNTs in ceramic compositions. The samples sintered by spark plasma sintering showed a finer grain size microstructure in the presence of CNTs compared to base samples. Mechanical and thermal properties show an enhancement with the addition of CNTs in the matrix. Raman Spectroscopy and high resolution transmission electron microscopy confirms CNT retention in the sintered nanocomposite.

INTRODUCTION

Aluminum Nitride (AlN) has attracted attention in recent years due to its high thermal conductivity, good mechanical strength, low-density and a low thermal expansion coefficient. Its resistance to molten metals makes it an ideal alternative material to alumina and beryllia currently used in microelectronics. AlN is attractive for applications in metal handling, as heat sinks, in semiconductor devices and electronic substrates, and as grinding media, seals, and filler materials. AlN is relatively brittle like all other ceramics, and therefore particulates such as Silicon Carbide [1; 11; 12; 17; 18; 19; 20; 23; 30; 32], Boron Nitride [2; 4], Titanium Diboride [19; 20; 36; 37], Titanium Nitride [9; 38] and even Zirconia[33] have also been used to increase its fracture toughness. Pure AlN is difficult to sinter to full density and therefore sintering additives – rare earth oxides of Yttrium[1; 3; 15; 27; 28; 29], Samarium[14; 34], Lanthanum[21], Cerium, along with fluorides of Calcium and Yttria[25] and even Molybdenum Disilicide[18] – have been used to enhance the sinterability of AlN[5]. AlN has been fabricated by various means: hot pressing[8], pressureless sintering, microwave sintering[10], SHS[7], reaction sintering[9; 16; 17; 22; 36] and more recently spark plasma sintering [6; 14; 24; 26; 31]. Based on previous research the addition of Silicon Carbide to the AlN matrix has been the most significant in improving the mechanical properties with a reported increase in fracture toughness up to 4.2 $MPa.m^{1/2}$ obtained by Tangen et al.[30] and a flexural strength of up to ~360 MPa is reported for Silicon Carbide reinforced Aluminum Nitride[17; 30].

Carbon nanotubes (CNTs) with elastic modulus greater than 1TPa and strength 10-100 times that of steel have been an attractive candidate for fundamental research since their discovery by Iijima[36]. Theoretical and experimental results have shown extremely high elastic modulus, greater than 1 TPa (the elastic modulus of diamond is 1.2 TPa)[37-41], and reported strengths 10–100 times higher than the strongest steel at a fraction of its weight. In addition to their extraordinary mechanical properties, carbon nanotubes also possess superior thermal properties. The CNTs are thermally stable up to 2800°C in vacuum, and the thermal conductivity is about twice as high as diamond[42-43]. CNTs with their outstanding mechanical properties and extraordinarily high aspect ratios are therefore ideal reinforcing fibers for ceramic matrix composites.[44-45]

In this research, Aluminum Nitride-CNT composites are fabricated by direct in-situ growth of CNTs on the AlN particles followed by densification via the Spark Plasma Sintering (SPS) technique.

EXPERIMENTAL DETAILS

Powder Mixtures

H.C. Starck Grade C Aluminum Nitride (average particle size = 0.8 to 1.8 microns with less than 2% oxygen) and Yttria (Inframat Advanced materials – average particle size = 0.15 microns) was mixed in a 97/3 wt% ratio, respectively. The powders were mixed using a rolling mill for 48 hours in baffled jars and then milled using silicon nitride balls in a rolling mill for 48 hours. The powders were mixed in ethanol in a Turbula shaker before processing (either sintering or catalyst coating) for 15 minutes and then dried, crushed and sieved before use.

In-Situ Growth of Multi-walled Nanotubes (MWNTs)

Previous research has shown that cobalt gives the highest yield of nanotubes per wt% addition[52]. CNTs can be grown by using $Co(NO_3)\cdot 6H_2O$ as a catalyst precursor[52,53]. $Co(NO_3)\cdot 6H_2O$ (Cobalt(II) nitrate hexadydrate, 98+%, A.C.S. reagent, Sigma-Aldrich) and AlN + 3 % Y_2O_3 powder (HC Starck) were mixed in ethanol, followed by sonication for 15 min. For CNT synthesis, AlN powder containing three different cobalt contents (Co/AlN = 0.5%, 1% and 2% by weight) was placed in a tube furnace fitted with a stirrer for all around growth of CNTs on the AlN particles by CVD technique. EDS scanning was performed to check for the distribution of the cobalt catalyst on the AlN particles especially for mixtures with 0.5 wt% catalyst.

CNT Yield Characterization

The CVD grown powders were analyzed using a field emission gun scanning electron microscope (FEGSEM - JEOL JSM-633OF), and X-ray diffraction (XRD). XRD was performed using a Siemens D5000 Diffraktometer employing Cu Ka radiation operated at 40 kV and 40 mA and fitted with a graphite monochromator. The step size and scan rate were fixed at 0.01° and 1sec/step, respectively. The yield of the in-situ growth of CNTs on AlN mixtures was characterized using TGA (TA Instruments: High–Res TGA 2950 Thermogravimetric Analyzer). The samples were heated in an oxidizing atmosphere with a heating rate of 20°C/min in the temperature range 50–800°C under 20ml/min ambient air, and the weight change was recorded.

Spark Plasma Sintering (SPS)
Spark plasma sintering of the powders was carried out using equipment from FCT System GmbH (Rauenstein, Germany) allowing several modes of operation (constant and pulsed DC current, pulse and pause times between 1 and 255 ms, different atmospheres: vacuum, N_2, Ar, high pressure forces: up to 250 kN). The base and in-situ CNT grown AlN-mixtures were placed in a 30 mm diameter graphite die and sintered under vacuum in the SPS unit.

Material Characterization
Microstructure of the sintered samples was analyzed using the FEGSEM. The density of the sintered samples was measured by the Archimedes method with distilled water as the immersion medium. XRD analysis was carried out for the powders before and after sintering to determine the chemical composition of each nanocomposite. The fractured surface of the samples was also analyzed using the FEGSEM.

CNT Retention
CNT composites were characterized using Raman Spectroscopy to ensure the CNT retention in the composites after sintering. The SPS sintered samples were fractured and Raman spectroscopy was carried out on the fractured surface. The Raman spectroscopy measurements were conducted at room temperature with a Raman spectrometer in the back scattering configuration. An argon ion laser tuned to 514 nm was used to collect the signals. To avoid a heating effect, the laser power was set at 3 mw after filter to excite the sample. Raman spectra were collected by using a high throughput holographic imaging spectrograph with a volume transmission grating, holographic notch filter, and thermoelectrically cooled charge coupled device (CCD) detector (Spectra Physics) with a resolution of 4 cm^{-1}. A 15-min exposure was used for each spectral collection.

Vickers Hardness Testing and 3 Point Bending
Vickers hardness testing was performed using a Wilson Tukon Hardness tester according to ASTM C1327 standard at loads of 300gf and 1000gf with a dwell time of 15 seconds. Five indentations were taken, and the average number was reported. The fracture toughness was characterized by measuring the length of the cracks formed at 1000 gf of loading using the FEGSEM. The fracture toughness was calculated using the following equation,

$$K_{IC} = 0.026\left(\frac{E^{1/2}P^{1/2}a}{C^{3/2}}\right) \qquad (1)$$

where P is the indentation load, "a" the half length of the indent, "C" the half length of the crack and E the elastic modulus of the composite. E is calculated from the elastic moduli of the components according the rule of mixtures.

Three-point bending tests were carried out on samples according to ASTM C1161 standard at ambient temperature using a Bose Enduratec machine. The crosshead speeds are maintained using displacement control, and the force was measured with a load cell accurate up to 0.1N.

Thermal Conductivity Measurement
Thermal diffusivity was measured using a laser flash technique on a Netzsch LFA-457 instrument on 10 mm x 10 mm x 2-4 mm samples in steps of 200°C from room temperature to 800°C, in an argon

atmosphere. The final thermal conductivity of the AlN-CNT composites was calculated automatically in the Netzsch instrument.

RESULTS AND DISCUSSION

Table 1 below shows the results of the TGA studies, density measurement, hardness testing, fracture toughness measurement and bending strength at room temperature.

Table 1: Vickers Hardness, Fracture Toughness and Three-point Bending Strength of the Samples

Sample ID	Composition (wt %)		Wt% CNT	% Theoretical Density	Hardness HV1/15 (GPa)	Fracture Toughness (Indentation) ($MPa.m^{1/2}$)	Bending Strength (MPa)
	AlN	Y_2O_3					
AlN - 3Y	97	3	0	99.27	12.89	2.8	411.67
A-0.5%Co	97	3	~7.4	92.34	10.69	3.6	524.10
A-1%Co	97	3	~16	86.23	8.93	3.9	363.29

EDS Scans for Cobalt Content

EDS scans were carried out on samples with 0.5 wt% catalyst to check for the uniformity of the catalyst distribution. Figure 1 below shows that Cobalt is distributed uniformly on the surface of the powders.

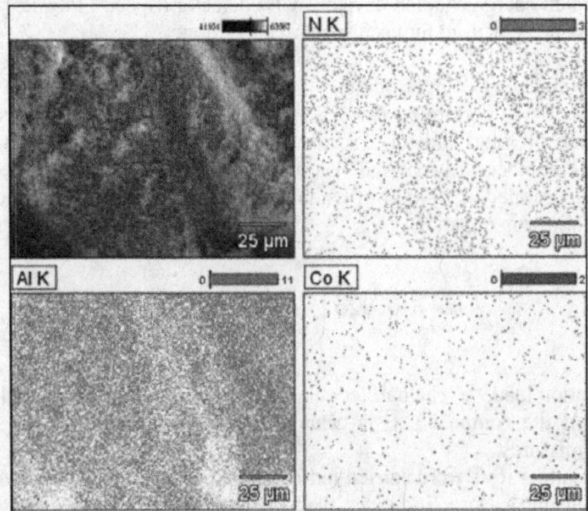

Figure 1: EDS scans for distribution of Cobalt in the AlN mixtures

FEGSEM, HRTEM, XRD and TGA Analysis of In-situ Grown CNTs on Aluminum Nitride

Figure 1 shows the FESEM images of in-situ grown CNTs on AlN particles after the CVD process. The CNTs can be seen covering the entire surface of the particles even at 2 wt% Co. The CNTs grown are MWNTs with an average diameter of 10-40nm and length in the range of 1- 5 microns. HRTEM imaging in Figure 2 shows substantial amount of MWNTs growing on the surface of the AlN particles.

Since the CVD process has established for tip growth mechanism, it is seen from the HRTEM images that most CNTs are attached to the particle without the presence of catalyst and hence the catalyst may possibly be removed by acid treatment later.

Figure 2: FESEM image of CNT on AlN for 2% Co

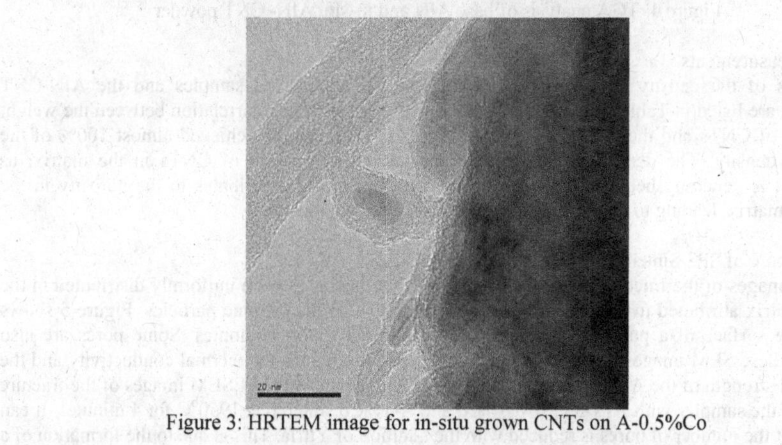
Figure 3: HRTEM image for in-situ grown CNTs on A-0.5%Co

The TGA curves of the as-received AlN base powder and Aluminum Nitride-CNT with 1 wt% and 2 wt% catalysts are illustrated in Figure 4 for comparison. Previous research by Huang et al.[56] has shown that MWNTs generally start oxidizing at approximately 400°C in air, and at 600°C rapid oxidation could take place[56]. Huang et al. also showed that amorphous carbon present on the surface of the Therefore, for this reason, all in-situ grown CNT powders were heated for 30 minutes at 300°C in air to remove the amorphous carbon. The weight loss between 400°C and 600°C is taken as the yield of the CNTs from the CVD growth process. It can readily be seen from the TGA analysis that even a small amount of catalyst can yield a great amount of CNT for the AlN particles. The yield of CNTs is ~ 16 wt% with 1 wt% catalyst and 22.5 wt% CNTs with 2 wt% catalyst. Other researchers have shown that

CNT-based composites lose their material integrity, such as hardness and strength, when the CNT weight percentage exceeds 10%, due primarily to poor sintering. Therefore, samples with only 1 wt% and 0.5 wt% catalyst, which gave about 16 wt% and 7.8 wt%, respectively, were designed and fabricated.

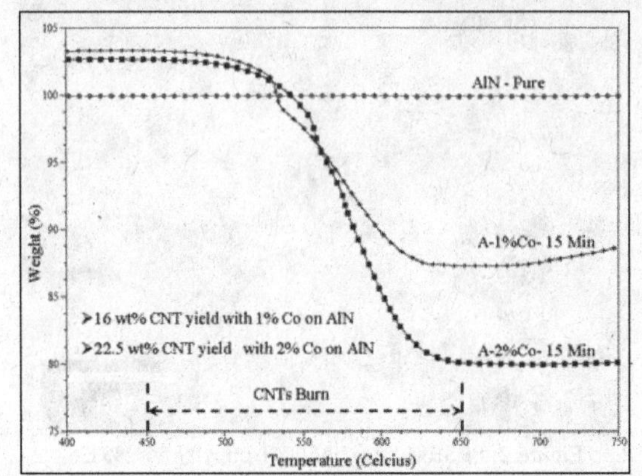

Figure 4: TGA analysis of base AlN and in-situ AlN- CNT powder

Density Measurements

The results of the density measurements of the sintered as-received samples and the AlN-CNT composites are listed in Table 1. It can be observed that there is a clear correlation between the weight percentage of CNTs and the density. The as-received AlN-3Y sample achieved almost 100% of the theoretical density. The density decreases with the increasing amount of CNTs in the matrix, as expected. It is generally believed that the hollow nature of CNTs contributes to the porosity in the composite matrix, leading to the diminishing hardness of the composites.

Microstructure of SPS Sintered Samples Uusing FESEM

FEFSEM images of the fracture surfaces demonstrate that the CNTs were uniformly distributed in the ceramic matrix attributed to the in-situ growth of the CNTs on the ceramic particles. Figure 5 shows the fracture surface of a pure AlN sample sintered at 1950°C for 4 minutes. Some pores are also evident in these SEM images. These pores could significantly reduce the thermal conductivity and the mechanical strength of the AlN composites. Figures 6, 7 and 8 show the FESEM images of the fracture surfaces of the samples with AlN-3Y, A-0.5%Co and A-1%Co sintered at 1950°C for 4 minutes. It can be seen that the number of pores is reduced with the addition of Yttria. This is due to the formation of a Yttria-Alumina eutectic phase at the grain boundary of the Aluminum Nitride. Since Alumina has very low thermal conductivity (30W/mk) compared to Aluminum Nitride, Yttria helps to increase the thermal conductivity of bulk ceramic by forming aluminate secondary phases. These phases are located around the AlN grains along the grain edge[13]. Along with enhancing the thermal conductivity, these aluminate phases formed by Yttria help in reducing the pores and thus increasing the mechanical strength of the ceramic composite. The SEM images illustrated in Figure 7 and 8 confirm the retention of the CNTs during the SPS process even at 1950°C. It can be seen from the SEM images that these samples exhibit very low porosity, implying that the CNTs may have bonded effectively with the AlN powders forming a dense ceramic matrix.

Figure 5: FESEM of etched surface of pure AlN sintered at 1950°C for 4 minutes

The fracture surfaces in Figure 7 and 8 demonstrate that CNTs are tightly embedded in the matrix. The fracture of the AlN-3Y samples generally occurs intergranularly, as evident from Figure 6, while the fracture of AlN-CNT composites, as shown in Figure 7 and Figure 8, appear to be a mixed mode of intergranular and transgranular features. The CNTs can be seen either at the grain boundaries or within the grain. It is also noticed in Figures 7 and 8 that the CNTs appear to have an effect of retarding the grain growth rate of the sintered composite. In addition to the strengthening effect, the grain-growth inhibition of CNTs may have a quite significant impact on the fabrication of CNT composites. It implies that CNTs could play a two-fold role in the nanocomposite: (1) enforcement and (2) prevention of grain growth, which could lead to a fine-grain composite.

Figure 6: Fractured surface of AlN-3Y-1950

Figure 7: Fractured surface of A-0.5%Co-1950

Figure 8: Fractured surface of A-1%Co-1950

XRD Analysis

XRD analysis of samples post SPS sintering, as shown in Figure 9 below, confirms that the sintered samples are predominantly single phase AlN. Yttria was not detected in the XRD scans due to its low percentage in the matrix. XRD scans also show that there are no unwanted reactions between the CNTs and the AlN matrix.

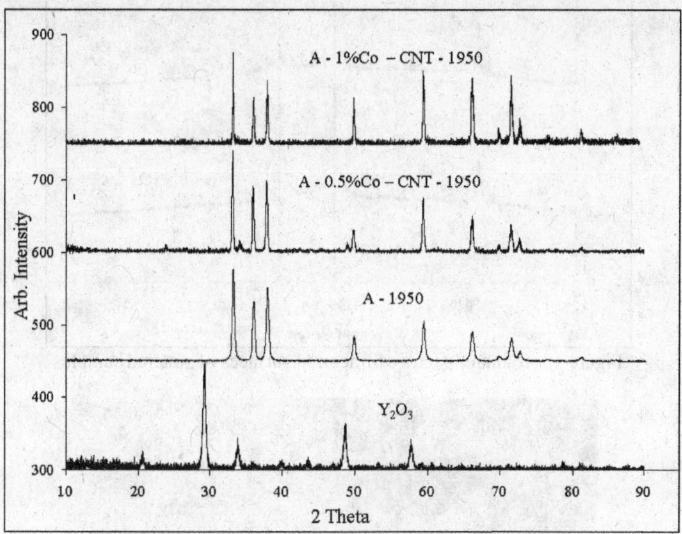

Figure 9: XRD analysis of samples after sintering

Raman Spectroscopic Analysis

Raman studies were conducted on the fracture surface of the sintered samples, as shown in Figure 10, to confirm the presence of CNTs after sintering. The analysis of the peaks from the Raman Spectra suggests that AlN retains its form after sintering and does not form other compounds. The Raman vibrational modes from the CNTs observed at 1373 cm-1 (D band) and 1565 cm-1 (G band) correspond to the MWNT modes reported in the literature[57]. Furthermore, the intensities and line profiles also suggest that these modes are from MWNTs and not from the other forms of carbon[58]. This observation thus suggests that the CNTs are retained after the SPS process.

HRTEM Studies Post Sintering

CNT retention is further confirmed by the HRTEM studies on the samples sintered at 1950°C. As seen from Figure 11, CNTs are observed in the sintered sample. To a certain extent, the CNTs appear to be damaged due to the nature of the SPS process. However, the damage seems to be confined to the outer walls, and the inner layers of CNTs appear to be intact.

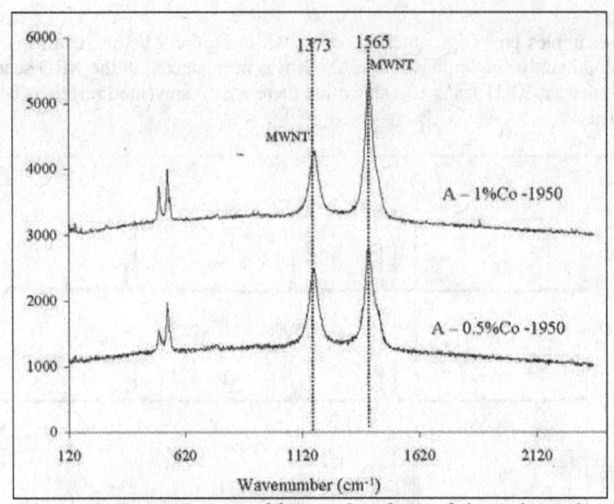

Figure 10: Raman Spectra of fractured surfaces of sintered samples

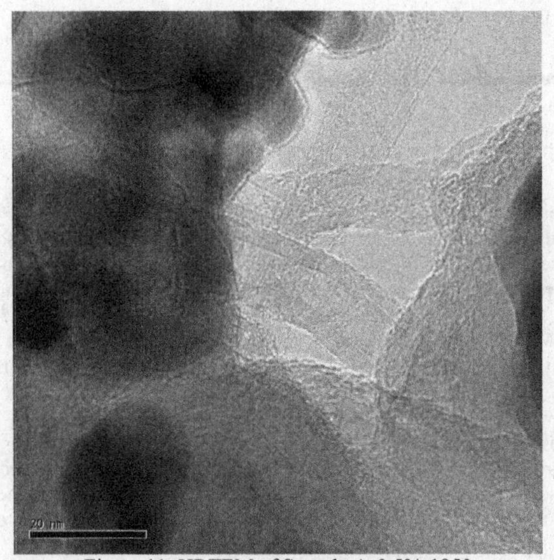

Figure 11: HRTEM of Sample A-0.5%-1950

Hardness and Fracture Toughness by Vickers Indentation Method

The testing results, as illustrated in Figure 12 below, indicates that the AlN-3Y sample demonstrates the highest Vickers Hardness value of 12.89 GPa, close to values reported in literature[35]. The Vickers hardness decreases with the increasing amount of CNTs in the matrix due to the porosity induced by the CNTs. Sample A-0.5% Co, which has about 7 wt% CNTs in the matrix, has a hardness of 10.69 GPa, while the sample with 16% CNTs has a 30% decrease in hardness when compared to the as-received AlN-3Y sintered at the same temperature for the same time. Fracture toughness calculated from the Vickers Indentations shows an increased value with the addition of CNTs in the matrix, as expected. A transgranular mode of fracture (seen in CNT reinforced AlN samples Figure 7 and Figure 8) consumes more energy than intergranular mode (Figure 6) and generally leads to higher fracture strength. The fracture toughness of A-1% Co sample increases approximately 1.25 times to that of the as-received AlN-3Y sample. The flexural strength of the base sample (AlN-3Y) is ~411 MPa higher than those obtained by Yijun et al. [35] The flexural strength of the sample obtained from the three-point bending test shows an interesting trend, with the flexural strength increasing by 25% with 7-wt% CNTs samples, and then reducing by 10% with 16-wt% CNTs samples, when compared to the as-received AlN-3Y samples. This can be attributed to the presence of too many CNTs in the matrix (16% wt), which weakens the composite. This trend is also clearly reflected in the density measurement of the samples where the sample AlN – 1% Co attains only approximately 85% theoretical density.

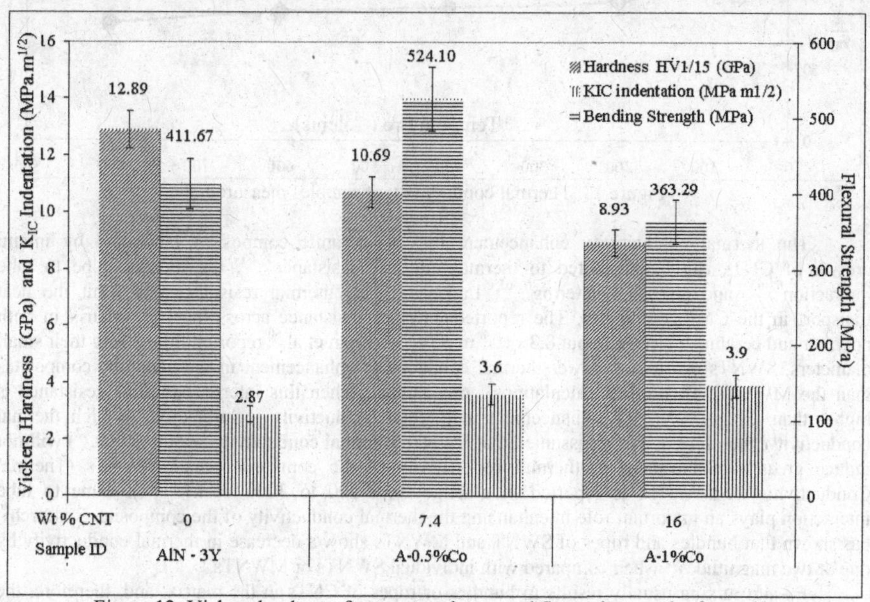

Figure 12: Vickers hardness, fracture toughness and flexural strength of samples

Thermal Conductivity Measurement

As the results in Figure 13 show, the thermal conductivity of the composite increases with the addition of CNTs in the matrix. Previous studies have shown that AlN ceramics are very sensitive to porosity[34-35]. However, in this study, the results show that the thermal conductivity increases substantially even with an increase in the amount of CNT, which may lead to more porosity in the matrix. The thermal conductivity at room temperature for A-1% Co is more than 2.5 times that of the as-received AlN-3Y. The sample A-0.5% Co also shows a 1.75-fold increase over the as-received sample.

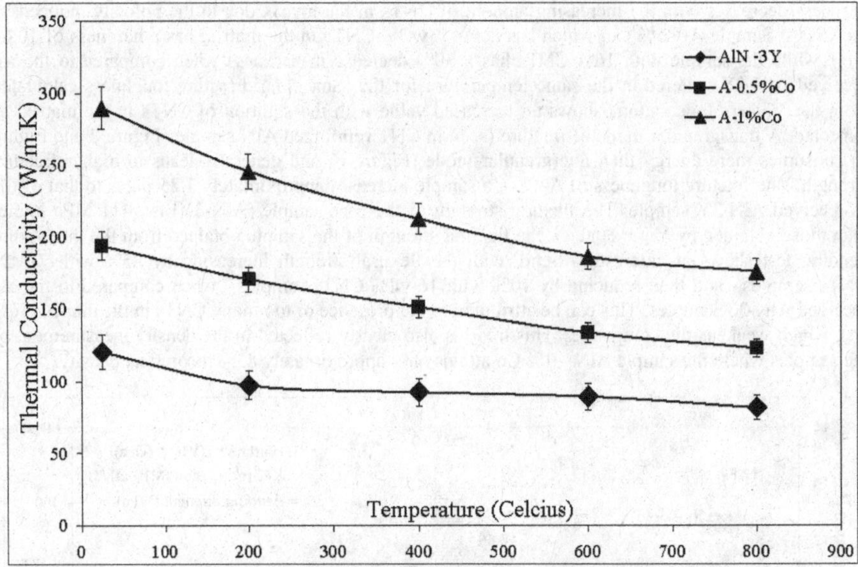

Figure 13: Thermal conductivity of samples measured

The thermal conductivity enhancement in CNT-ceramic composites fabricated by in-situ growth of CNTs can be attributed to thermal interface resistance[59-63], the reduced tube to tube interaction[62-63], and phonon scattering[64-65]. Large interface thermal resistance can limit the heat transport in the CNT composites. The reported interface resistance across the CNT-matrix in both polymer and ceramic matrix is about 8.3×10^{-8} m^2K/W[59; 60]. Nan et al.[60] reported that, due to their small diameters, SWNTs can induce lower thermal conductivity enhancement in the nanotube composites than the MWNTs. Theoretical calculations[59] predict that, when this interface thermal resistance is higher than 10^{-8} m^2K/W, the enhancement in thermal conductivity is insensitive to high thermal conductivity; therefore, it can be assumed that SWNTs (thermal conductivity ~6600 W/mK[62]) will not induce greater enhancement in thermal conductivity of the composites than MWNTs (Thermal Conductivity for MWNTs is reported in a range from 200 to 3000 W/mK[62-63].). Tube to tube interaction plays an important role in enhancing the thermal conductivity of the composite. Research[60] has shown that bundles and ropes of SWNT and MWNTs show a decrease in thermal conductivity by one or two magnitudes[62] when compared with individual SWNTs or MWNTs.

Ex-situ mixing usually results in bundles or ropes of CNTs in the matrix, and, therefore, the resulting inter-tube phonon scattering can affect thermal conductivity of the composite as a whole. Since phonon transport dominates the thermal conductivity for SWNT and MWNT at all temperatures, the enhancement in thermal conductivity of the composites would be dependent predominantly on the phonon transport along the nanotubes. Intertube phonon transport will also degrade the longitudinal thermal conductivity of the nanotubes[67]. In our work, most of the MWNTs are single tubes; therefore, the heat loss due to thermal diffusion between the individual tubes is reduced when compared to bundles of one or more nanotubes as seen in ex-situ mixing.

CONCLUSIONS

1. In this research, multi-walled carbon nanotubes have been directly grown in-situ on Aluminum Nitride powders. The grown MWNTs have been characterized using FEGSEM, and the yield with different catalyst ratios has been analyzed using the TGA. The Aluminum Nitride powders with the in-situ grown nanotubes were then sintered using spark plasma sintering. The sintered samples were analyzed for mechanical and thermal properties. This study shows that using the in-situ CVD technique can result in a uniform distribution of CNTs in the ceramic matrix. Our studies also demonstrate that CNTs can be retained after the SPS process.
2. The mechanical testing shows an increase in flexural strength and in thermal conductivity of the composite fabricated from Aluminum Nitride-CNT powders when compared to base AlN-3Y samples. The AlN-CNT with 16 wt% CNT composites can have up to 2.5 times increase in thermal conductivity when compared with the pure AlN ceramic. The optimum properties are obtained with 0.5% catalyst, which results in approximately 7 wt% CNTs in the ceramic matrix. This sample A-0.5% Co shows an enhancement of the thermal conductivity, the flexural strength and the fracture toughness with no appreciable decrease in hardness.
3. The method developed in this research proves to be able to distribute CNT much more uniformly compared to the conventional ex-situ mixing method and can be extend to mass production for commercial applications.
4. In addition to the strengthening effect, through this study, it is observed that CNTs can also minimize the grain growth in the AlN-CNT composites. This important feature clearly indicates that CNT plays triple roles in the CNT-ceramic nanocomposites: (1) strengthening effect, (2) enhancing the thermal conductivity and (3) inhibition of grain growth. More benefits of CNTs composites are to be explored.
5. Average grain size of the sintered composites showed an inverse relation with the CNT content, which can be used to tailor the composite according to application needs.

REFERENCES

[1]R. M. Balestra, S. Ribeiro, S. P. Taguchi, F. V. Motta, andC. Bormio-Nunes, "Wetting behaviour of Y2O3/AlN additive on SiC ceramics," Journal of the European Ceramic Society, 26[16] 3881-86 (2006).

[2]A. A. Buchheit, G. E. Hilmas, W. G. Fahrenholtz, D. M. Deason, andH. Wang, "Mechanical and thermal properties of AlN-BN-SiC ceramics," *Materials Science and Engineering: A,* 494[1-2] 239-46 (2008).

[3]H. Buhr and G. Müller, "Microstructure and thermal conductivity of AlN(Y2O3) ceramics sintered in different atmospheres," *Journal of the European Ceramic Society,* 12[4] 271-77 (1993).

[4]W. S. Cho, Z. H. Piao, K. J. Lee, Y. C. Yoo, J. H. Lee, M. W. Cho, Y. C. Hong, K. Park, andW. S. Hwang, "Microstructure and mechanical properties of AlN-hBN based machinable ceramics prepared by pressureless sintering," *Journal of the European Ceramic Society,* 27[2-3] 1425-30 (2007).

[5]X. Du, M. Qin, A. Farid, I. S. Humail, andX. Qu, "Study of rare-earth oxide sintering aid systems for AlN ceramics," *Materials Science and Engineering: A,* 460-461 471-74 (2007).

[6]X. Du, M. Qin, A. Rauf, Z. Yuan, B. Yang, andX. Qu, "Structure and properties of AlN ceramics prepared with spark plasma sintering of ultra-fine powders," *Materials Science and Engineering: A,* 496[1-2] 269-72 (2008).

[7]R. Fu, K. Chen, X. Xu, andJ. M. F. Ferreira, "Highly crystalline AlN particles synthesized by SHS method," *Materials Letters,* 59[19-20] 2605-09 (2005).

[8]W. Han, G. Li, X. Zhang, andJ, Han, "Effect of AlN as sintering aid on hot-pressed ZrB2-SiC ceramic composite," *Journal of Alloys and Compounds,* In Press, Corrected Proof.

[9]G. Hongyu, Y. Yansheng, L. Aiju, L. Yingcai, Z. Yuhua, andL. Chunsheng, "Reaction sintering fabrication of (AlN, TiN)-Al2O3 composite," *Materials Research Bulletin,* 37[9] 1603-11 (2002).

[10]C.-Y. Hsieh, C.-N. Lin, S.-L. Chung, J. Cheng, andD. K. Agrawal, "Microwave sintering of AlN powder synthesized by a SHS method," *Journal of the European Ceramic Society,* 27[1] 343-50 (2007).

[11]J. Ihle, M. Herrmann, andJ. Adler, "Phase formation in porous liquid phase sintered silicon carbide: Part I:: Interaction between Al2O3 and SiC," *Journal of the European Ceramic Society,* 25[7] 987-95 (2005).

[12]J. Ihle, M. Herrmann, andJ. Adler, "Phase formation in porous liquid phase sintered silicon carbide: Part III: Interaction between Al2O3-Y2O3 and SiC," *Journal of the European Ceramic Society,* 25[7] 1005-13 (2005).

[13]J. Jarrige, J. P. Lecompte, J. Mullot, andG. Müller, "Effect of oxygen on the thermal conductivity of aluminium nitride ceramics," *Journal of the European Ceramic Society,* 17[15-16] 1891-95 (1997).

[14]K. A. Khor, L. G. Yu, andY. Murakoshi, "Spark plasma sintering of Sm2O3-doped aluminum nitride," *Journal of the European Ceramic Society,* 25[7] 1057-65 (2005).

[15]Y.-W. Kim, Y.-S. Chun, T. Nishimura, M. Mitomo, andY.-H. Lee, "High-temperature strength of silicon carbide ceramics sintered with rare-earth oxide and aluminum nitride," *Acta Materialia,* 55[2] 727-36 (2007).

[16]Y. W. Kim, H. C. Park, Y. B. Lee, K. D. Oh, andR. Stevens, "Reaction sintering and microstructural development in the system Al2O3-AlN," *Journal of the European Ceramic Society,* 21[13] 2383-91 (2001).

[17]Y. Kobayashi, M. Sugimori, J. F. Li, A. Kawasaki, R. Watanabe, S. Somiya, R. P. H. Chang, M. Doyama, andR. Roy, "Reaction Synthesis of SiC-AlN Ceramic Alloys," pp. 289-92. in Materials Science and Engineering Serving Society. Elsevier Science B.V., Amsterdam, 1998.

[18]K. Krnel, D. Sciti, E. Landi, andA. Bellosi, "Surface modification and oxidation kinetics of hot-pressed AlN-SiC-MoSi2 electroconductive ceramic composite," *Applied Surface Science,* 210[3-4] 274-85 (2003).

[19]V. A. Lavrenko, J. Desmaison, A. D. Panasyuk, M. Desmaison-Brut, andE. Fenard, "High-temperature oxidation of AlN-SiC-TiB2 ceramics in air," *Journal of the European Ceramic Society,* 25[10] 1781-87 (2005).

[20]S.-H. Lee, S. Guo, H. Tanaka, K. Kurashima, T. Nishimura, andY. Kagawa, "Thermal decomposition, densification and mechanical properties of AlN-SiC(-TiB2) systems with and without B, B4C and C additives," *Journal of the European Ceramic Society,* 28[8] 1715-22 (2008).

[21]X. L. Li, H. A. Ma, Y. J. Zheng, Y. Liu, G. H. Zuo, W. Q. Liu, J. G. Li, andX. Jia, "AlN ceramics prepared by high-pressure sintering with La2O3 as a sintering aid," *Journal of Alloys and Compounds,* 463[1-2] 412-16 (2008).

[22]A. Maghsoudipour, M. A. Bahrevar, J. G. Heinrich, andF. Moztarzadeh, "Reaction sintering of AlN-AlON composites," *Journal of the European Ceramic Society,* 25[7] 1067-72 (2005).

[23]G. Magnani and L. Beaulardi, "Long term oxidation behaviour of liquid phase pressureless sintered SiC-AlN ceramics obtained without powder bed," *Journal of the European Ceramic Society,* 26[15] 3407-13 (2006).

[24]M. Omori, "Sintering, consolidation, reaction and crystal growth by the spark plasma system (SPS)," *Materials Science and Engineering A,* 287[2] 183-88 (2000).

[25]L. Qiao, H. Zhou, andR. Fu, "Thermal conductivity of AlN ceramics sintered with CaF2 and YF3," *Ceramics International,* 29[8] 893-96 (2003).

[26]L. Qiao, H. Zhou, andC. Li, "Microstructure and thermal conductivity of spark plasma sintering AlN ceramics," *Materials Science and Engineering B,* 99[1-3] 102-05 (2003).

[27]L. Qiao, H. Zhou, H. Xue, andS. Wang, "Effect of Y2O3 on low temperature sintering and thermal conductivity of AlN ceramics," *Journal of the European Ceramic Society,* 23[1] 61-67 (2003).

[28]K. Suzuki and M. Sasaki, "Microstructure and mechanical properties of liquid-phase-sintered SiC with AlN and Y2O3 additions," *Ceramics International,* 31[5] 749-55 (2005).

[29]T. S. Suzuki and Y. Sakka, "Preparation of oriented bulk 5 wt% Y2O3-AlN ceramics by slip casting in a high magnetic field and sintering," *Scripta Materialia,* 52[7] 583-86 (2005).
[30]I.-L. Tangen, Y. Yu, T. Grande, T. Mokkelbost, R. Høier, andM.-A. Einarsrud, "Preparation and characterisation of aluminium nitride-silicon carbide composites," *Ceramics International,* 30[6] 931-38 (2004).
[31]D. Tiwari, B. Basu, andK. Biswas, "Simulation of thermal and electric field evolution during spark plasma sintering," *Ceramics International,* 35[2] 699-708 (2009).
[32]H. A. Toutanji, D. Friel, T. El-Korchi, R. N. Katz, G. Wechsler, andW. Rafaniello, "Room temperature tensile and flexural strength of ceramics in AlN-SiC system," *Journal of the European Ceramic Society,* 15[5] 425-34 (1995).
[33]C. Toy and E. Savrun, "Novel composites in the aluminum nitride-zirconia and --hafnia systems," *Journal of the European Ceramic Society,* 18[1] 23-29 (1998).
[34]X. Xu, H. Zhuang, W. Li, S. Xu, B. Zhang, andX. Fu, "Improving thermal conductivity of Sm2O3-doped AlN ceramics by changing sintering conditions," *Materials Science and Engineering A,* 342[1-2] 104-08 (2003).
[35]Y. Yijun and Q. Tai, "Effects of Behaviors of Aluminum Nitride Ceramics with Rare Earth Oxide Additives," *Journal of Rare Earths,* 25[Supplement 1] 58-63 (2007).
[36]. Iijima S. Helical microtubules of graphitic carbon. Nature 1991; 354:56–8.
[37]. A. Peigney, Ch. Laurent, E. Flahaut, and A. Rousset Carbon nanotubes in novel ceramic matrix nanocomposites. Ceram Int 2000; 26:677–83.
[38]. G. Van Lier, C. Van Alsenoy, V. Van Doren, and P. Geerlings, Ab initio study of the elastic properties of single-walled carbon nanotubes and graphene. Chem Phys Lett 2000; 326:181–5.
[39]. M.M.J. Treacy, T.W. Ebbesen, and J.M. Gibson, Exceptionally high Young's modulus observed for individual carbon nanotubes. Nature 1996; 381:678–80.
[40]. M.F. Yu, O. Lourie, M.J. Dyer, K. Moloni, T.F. Kelly, and R.S. Ruoff, Strength and breaking mechanism of multiwalled carbon nanotubes under tensile load. Science 2000; 287:637–40.
[41]. A. Thess, R. Lee, P. Nikolaev, H. Dai, P. Petit, J. Robert, C. Xu, Y.H. Lee, S.G. Kim, A.G. Rinzler, D.T. Colbert, G.E. Scuseria, D. Tomanek, J.E. Fischer,and R.E. Smalley, Crystalline ropes of metallic carbon nanotubes. Science 1996; 273:483–7.
[42]. Ando Y, Zhaoa X, Shimoyama H, Sakai G, and Kaneto K, Physical properties of multiwalled carbon nanotubes. Int J Inorg Mater 1999; 1:77–82.
[43]. M.J. Biercuk, M.C. Llaguno, M. Radosavlijevic, J.K. Hyun, and A.T. Johnson, Carbon nanotube composites for thermal management. Appl Phys Lett 2002; 80:2767–9.
[44]. R.H. Baughman, A.A. Zakhidov, and W.A. de Heer, Carbon nanotubes—the route toward applications. Science 2000; 297:787–92.
[45]. K.M. Prewo, Fiber-reinforced ceramics: new opportunities for composite materials. Am. Ceram Soc Bull 1989; 68:395–400.
[46]. J. Tatami, T. Katashima, K. Komeya, T. Meguro, and T. Wakihara, Electrically Conductive CNT-Dispersed Silicon Nitride Ceramics. Journal of American Ceramic Society 2005; 88 [10]: 2889–93.
[47]. S. Y. Lee, H. Kim, P. C. McIntyre, K. C. Saraswat, and J. S. Byun, Atomic Layer Deposition of ZrO_2 on W for Metal–Insulator–Metal Capacitor Application. Appl. Phys Letter 2003; 82:2874–6.
[48]. E.T. Thostenson, Z. Ren, and T.W. Chou, Advances in the science and technology of carbon nanotubes and their composites: a review. Compos Sci Technol 2001; 61:1899–912.
[49]. H. Dai, Carbon nanotubes: opportunities and challenges. Surf Sci 2002; 500:218–41.
[50]. K. Ahmad, W. Pan, and S.L. Shi. Electrical conductivity and dielectric properties of multiwalled carbon nanotube and alumina composites. Appl. Phys. Lett. 2006; 89: 133122.
[51]. G.D. Zhan, J.D. Kuntz, J.E. Garay, and A.K. Mukherjee, Electrical properties of nanoceramics reinforced with ropes of single-walled carbon nanotubes. Appl. Phys. Lett. 2003; 83: 1228.

[52] H.J. Dai, A.G. Rinzler, P. Nikolaev, A. Thess, D.T. Colbert, and R.E. Smalley, Single-wall nanotubes produced by metal-catalyzed disproportionation of carbon monoxide. Chem. Phys. Lett 1996; 260: 471.
[53] W.Z. Li, J.G. Wen, M. Sennett, and Z.F. Ren, Clean double-walled carbon nanotubes synthesized by CVD. Chem Phys Lett 2003; 368:299–306.
[54] Laurent Ch, Peigney A, Dumortier O, Rousset A. Carbon nanotubes–Fe–alumina nanocomposites. Part II: microstructure and mechanical properties of the hotpressed composites. J Eur Ceram Soc 1998; 18:2005–13.
[55] X. Wang, N.P. Padture, and H. Tanaka, Contact-damage-resistant ceramic/single-wall carbon nanotubes and ceramic/graphite composites. Nat Mater 2004; 3:539–44.
[56] W. Huang, Y. Wang, G. Luo, and F. Wei, 99.9% purity multi-walled carbon nanotubes by vacuum high-temperature annealing, Carbon 41 (2003) 2585–2590.
[57] A.P. Naumenko, N.I. Berezovska, M.M. Biley, and O.V. Shevchenko, Vibrational analysis and Raman spectra of tetragonal Zirconia, Physics and Chemistry of Solid State, 2008, 121 – 125.
[58] P.C. Eklund, J.M. Holden, and R.A. Jishi, Vibrational modes of carbon nanotubes; spectroscopy and theory. Carbon 1995, 959-972.
[59] Huxtable, S., Cahill, D.G., Shenogin, S., Xue, L., Ozisik, R., Barone, P., Usrey, M., Strano, M.S., Siddons, G., Shim, M., and Keblinski, P., Nat. Mater. **2**, 731(2003).
[60] Nan, C.W., Li, X.P., and Birringer, R., J. Am. Ceram. Soc. **83**, 848 (2000).
[61] Nan, C.W., Liu, G., Lin, Y.H., and Li, M., Appl. Phys. Lett. **85**, 3549 (2004).
[62] Biercuk, M.J., Llaguno, M.C., Radosavlijevic, M., Hyun, J.K., and Johnson, A.T., Appl. Phys. Lett. 80, 2767 (2002).
[63] Yang, D.J., Zhang, Q., Chen, G., Yoon, S.F., Ahn, J., Wang, S.G., Zhou, Q., Wang, Q., and Li, J.Q., Phys. Rev. B **66**, 165440 (2002).
[64] Kim, P., Shi, L., Majumdar, A., and McEuen, P.L., Phys. Rev. Lett. **87**, 215502 (2001).
[65] Ginnings, D.C., and Furukawa, G.T., J. Am. Chem. Soc. **75**, 522 (1953).
[66] Dresselhaus, M.S., Dresselhaus, G., and Avouris, P., Carbon Nanotubes: Synthesis, Structure, Properties, and Applications, Springer, Berlin, 2001.
[67] Hone, J., Whitney, M., Piskoti, C., and Zettle, A., Phys. Rev. B 59, R2514 (1999).
[68] Zhang, H.L., Li, J.F., Yao, K.F., and Chen, L.D., J. Appl. Phys. 97, 114310 (2005).

MICROSTRUCTURAL CHARACTERIZATION OF C/C-SIC COMPOSITES AFTER OXIDATION WITH OXYACETYLENE GAS IN OPEN ATMOSPHERE

V. K. Srivastava* and Shraddha Singh

*Professor in Production Engineering
Department of Mechanical Engineering
Institute of Technology
Banaras Hindu University, Varanasi – 221005, INDIA
E-mail: vk_sa@yahoo.co.in and vijays.sa55@gmail.com

ABSTRACT

C/C-SiC composites plates were directly exposed with an oxyacetylene gas welding flame to understand the effect of temperature (>1000°C) in open atmosphere. Duration of exposure time was increased to measure the weight loss and performance of composites. SEM and TG analysis were performed to see the microstructural behavior of C/C-SiC composites after the exposure of oxyacetylene gas flame in open atmosphere. The results show that the weight loss increases with increase of exposure time clearly indicating that the excess amount of carbon is oxidized to form carbon mono-oxide. The matrix materials undergo pyrolysis reactions within short periods of time generating volatiles and leaving behind a porous skeleton of carbon fibres. Scanning Electron Microscopy (SEM) fractograph of fractured samples indicates that fibres debonded, broken and pull out due to increase of applied load, because, excess amount of matrix oxidized with the effect of high temperature.

1. INTRODUCTION:

Ceramic matrix composites (CMC) are interesting materials for an increasing number of applications in aerospace industry as well as in civil engineering. This class of materials offers high strength to density ratios. Also, their higher temperature capability over conventional supperalloys may allow for components that require little or no cooling. This benefit can lead to simpler component design and weight savings. These materials can also contribute to increase in operating efficiency due to high operating temperatures being achieved. The wide range of applications for CMCs includes combustor linners, turbomachinery, aircraft brakes, nozzles and thrusters.

When evaluating CMCs for potential use in high temperature structural applications, the basic characterization of the material obtained from mechanical and environmental testing is important in understanding the fundamental properties of the materials. These types of tests however may not be able to provide enough information on how the material will perform at its applications. This is especially true, when many variables may include such as high temperature, mechanical and thermal stresses, flowing gases, reactive environments, high chamber pressure and material reactivity [1-4].

To meet the above criterion, C/C–SiC composites is one of the key materials, which can manufactured by liquid silicon infiltration (LSI) process, where porous carbon/carbon preforms are infiltrated with melted silicon to form silicon carbide. This process leads to damage-tolerant ceramics that have significantly lower component fabrication time and, therefore, reduced component costs compared to other CMCs manufacturing processes. Not only their properties of the reinforcement and the matrix but also the interfacial properties between them determine mechanical properties of composites. Especially, the interfacial shear strength is very important for the strength of brittle matrix composites such as C/C composite, and it is changed drastically by heat treatment [5-6]. However, their oxidation in oxidizing environment above 673 K limits their applications in high temperature. For the good quality of the fibre-matrix interface and constituent properties, microstructure is a key factor in determining engineering properties in this class of materials. Microstructural characteristics such as cracking, weight loss and porosity are found to correlate to properties such as tensile strength, flexural strength and elastic modulus [7]. The use of fractography in failure analysis is well established. It is often the case that fractographic evidence is crucial in correctly identifying the sequence of events in a failure, and in proving, beyond reasonable doubt, a particular scenario to be the most feasible. This applies equally to many cases involving polymers or ceramics, as well as to metals.

Despite the attractiveness of fiber- reinforced composites as engineering components, they are not currently being applied to the extent that they could be. Even when they have been employed, relatively low stress applications and large safety factors were usually considered. The main reason is the difficulty and uncertainty that exist in determining their failure strength, fracture toughness, operating lifetime in severe conditions because the nature of the deformation and failure behavior of the composites were very complicated [9-10].

This paper examined the fracture morphology and effect of temperature on C/C-SiC composites with the help of SEM and TGA methods. Also develop the understanding of oxidation effect on fibre and matrix interfaces.

Experimentation:

Material:

Carbon- carbon fibre reinforced silicon carbide (C/C-SiC) is a candidate material to be used in industrial applications. C/C SiC was obtained from MPA, Stuttgart, Germany. DLR, Stuttgart, Germany has developed C/C-SiC composites by LSI method, which is tough when the fibre matrix bonding is properly optimized through a thin layer of an interfacial material referred to as the interphase.

1.2. SEM analysis:

The microstructure of fractured sample was observed by a scanning electron microscope (SEM). The surface of fractured specimen to be examined is scanned with an electron beam, and then reflected beam of electron is collected, and then displayed at the same scanning rate on a cathode ray tube. The image on the screen, which may be photographed, represents the surface features of the specimen. The surface may or may not be polished and etched but it must be electrical conductive, a very thin metallic surface coating must be applied to non conductive materials. Microscopy examination is a useful tool in the characteristic of materials. This will become apparent in subsequent chapters that correlate the microstructure with various characteristics and properties. Microstructure examination also used to determine the mode of mechanical fracture, to predict the oxidation effect on mechanical property.

TG Analysis:

Perkin-Elmer Pyris Diamond TG/DTA investigates the thermal stability of C/C-SiC composites in the presence of nitrogen gas by heating the samples at the rate 750 K/min. The weight loss of oxidized C/C-SiC composite specimens was recorded with the variation of oxidation time.

2. RESULTS AND DISCUSSION

When material is exposed to oxy-acetylene gas torch it could degrade because of oxidation. Oxygen could attack the fibres especially in the vicinity of cracks or porosity. A chemical reaction could also occur in the matrix if non-reacted Si from the infiltration process is available. The lower viscosity of the SiO_2 at high temperatures could result in a shelf healing of micro cracks.

The microstructure indicates severe reaction between carbon fibre and molten silicon, which indicates that the microcracks appear more, pronounced as can be seen in Fig.1. This increase in defects is due to the morphological change in the sample. After oxidation carbon degraded in the form of carbon-di-oxide and carbon-mono-oxide ($C+O_2 \rightarrow CO_2\uparrow$, $2C+O_2 \rightarrow 2CO\uparrow$) and Si reduce in the form of silicon oxide ($Si+O_2 \rightarrow SiO_2$).

Thermal conductivity is a key property in many applications of C/C-SiC composites. Generally these are relatively good conductors of heat but their conductivity depends upon the crystallinity of their constituents. And due to change in the proportion of oxygen, carbon and silicon, the conductivity becomes also changed. Also, morphological observation appears that the mode of fracture surface was appeared under different loading conditions and matrix portion failed like brittle pattern as shown in Fig. 2(a). The SEM at higher magnification observed that the fibre debonded, pull-out and broken from different stages. The deformation of fibres, matrix and the distribution of constituent particle were quite different on microscopic scale due to the fracture. Fig. 2 (b & c) shows that the cleavage fractured appear due to deformation of carbon matrix at higher temperature. However, carbon matrix was fully oxidized and shrinkage was appeared around the fibre as can be seen in Fig. 2(d).

Fig.1. SEM fractograph of C/C-SiC composites (X 20 μm)

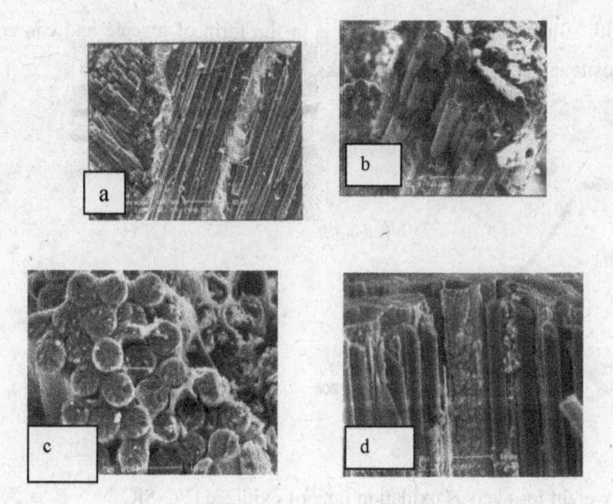

Fig.2. SEM micrographs of C/C-SiC oxidized at 800^0C for 10 min under tensile showing showing (a) delamination of fibres (X 60 μm) (b) broken fibre bundles (X 20 μm) (c) shear deformation of matrix along with reach area of SiC (X 10 μm) and (d) shrinkage/brittle fracture of SiC matrix around the fibres (X 10 μm).

3.2. TGA analysis

Before the effect of a constant load on C/C-SiC composites is evaluated, it is beneficial to have an understanding of how the material and its fibre constituent behave in unstressed, high temperature oxidizing conditions. The present weight loss of oxidized C/C-SiC composite versus oxidation time is shown in Fig. 3. Isothermal TGA oxidation tests was conducted to investigate the influence of temperature effect on the overall weight losses. Constant reaction rates are seen for oxidation time ranging from 0 to 150 minutes. Faster reaction rates and less of temperature dependence are observed at oxidation time ranging from 150 to 400 min.

As the fibre and matrix expand to their original length at the processing temperature, residual thermal stresses are relieved and the pre-existing microcracks become more pronounced and pitch off the supply of oxygen to the fibres. The second factor is due to the formation of silica as oxygen reacts with silicon carbide. The oxide is in the form of a scale and can cover the surface of the composite and also full in microcracks.

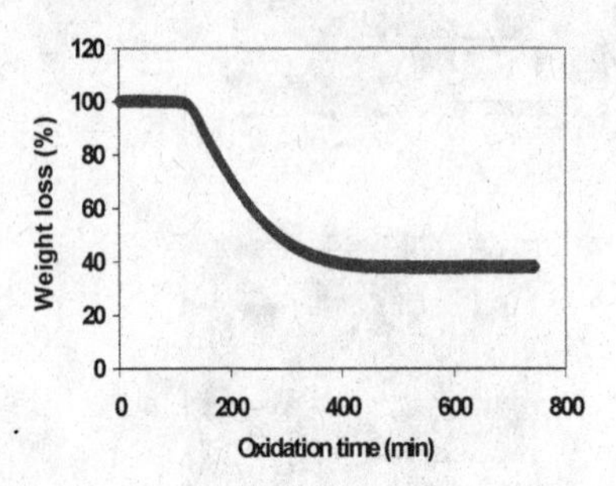

Fig. 3. Variation of weight loss versus oxidation time of oxidized C/C-SiC

After all cracks are closed the C/C–SiC substrate is completely protected against oxygen ingress, so no further mass losses occur. On the other hand, oxygen can reach the SiC layer only through cracks in the cordierite layer. Oxidation of SiC is thus limited and, as cordierite is chemically stable in oxidative atmospheres like most oxides, mass gains cannot be observed. At high temperature no oxidation of SiC occurs so that coating cracks are not sealed. Oxygen can thus enter the C/C–SiC substrate and lead to the observed high oxidation rates of carbon at high temperature.

4. REMARKS

Based on the morphological and TG observations, the results clearly indicate that weight loss of C/C-SiC composite is increased, which increase the debonding the fibre, fibre pullout and shrinkage of SiC matrix around the fibre as well as marix rich area.

REFERENCES:

[1] O. Yamamoto, K. Imai, T. Samoto, J. Eur. Ceram. Soc. 12 (1993) 435.
[2] Q. S. Zhu, X.L. Qiu, C.W. Ma, Carbon 37 (1999) 1475.
[3] S. Singh and V.K. Srivastava, Materials Science & Engg.-A, 486 (2008) 534-539.
[4] T.L. Dhami, O.P. Bahl, B.R. Awasthy, Carbon 33 (1995) 479.
[5] N. Iwashita, Y. Sawada, K. Shimizu, S. Shinke, H. Shioyama, Carbon 1995; 33:405-413.
[6] M.E. Westwood, J.D. Webstar, R.J. Day, F.H. Hays, R. Taylor, J. Mater. Sci. 31(1996) 1389.
[7] J.C. Williams, S.W. Yurgartis, J.C. Moosbrugger, J Compos Mater 1996; 30(7): 785-800.

[8] M.N.James, Wei LiWu. Hail, Damage to polycarbonate roofing sheets. In: Bicego V, et al., editors. Proceedings of an international conference on case histories on integrity and failure in industry, Milan, September 1999. Cradley Health, West Midlands: EMAS. 1999. pp. 715-24.
[9] M. Saki, J. Ceram. Soc. Jpn. 99 (10) (1991) 983- 992.
[10] A.G. Evans, J. Am. Ceram. Soc. 73 (1990) 187.

THE EFFECT OF INTERPARTICLE INTERACTIONS ON THE RHEOLOGICAL PROPERTIES OF PARAFFIN-WAX SUSPENSIONS

AlešDakskobler [*], TomaKosmač
Engineering Ceramics Department, Joef Stefan Institute,
Ljubljana, Slovenija

ABSTRACT

The overall success of the low-pressure injection moulding (LPIM) of ceramics relies primarily on the properties of the ceramic-powder paraffin-wax suspension. It is desirable that the suspension contains a high loading of homogeneously dispersed particles in a liquid carrier to ensure suitable moulding characteristics.

These suspensions are weakly flocculated because the surfactant, which is usually stearic acid, is insufficient to provide true stabilization, but is only effective in reducing the van der Waals attraction. When the overall interaction potential in the suspension is attractive, there is a certain volume fraction of solids whereby a continuous network is formed and a certain shear stress is needed to overcome the interparticle forces in order to induce flow and affect the flow characteristics of these suspensions. The suspensions used in LPIM are all well above this critical volume where a network of attractive particles is formed. This shows that the attractive interparticle interactions play a crucial role in obtaining as high as possible solids loading and the appropriate rheological characteristics of the paraffin-wax suspensions.

In this study we focus on the attractive interparticle interactions that occur in the alumina-powder, paraffin-wax suspensions and their effect on the rheological properties. The effects of the particle size of the powders and the solids loading on the rheological properties of paraffin-wax suspensions are presented.

INTRODUCTION

Low-pressure injection moulding (LPIM) is a cost-effective, near-net-shape method for producing complex-shaped ceramic components.[1,2,3] In this process, suspensions of a ceramic powder and paraffin wax are used to form green ceramic parts. These suspensions must have a low viscosity at the moulding temperature (65-70°C), at which they are injected into a cold, metallic model, where the suspension cools below its melting temperature and retains the shape of the model.

The success of the low-pressure injection moulding of ceramics relies primarily on the properties of the ceramic-powder paraffin-wax suspension. It is desirable that the suspension contains a high loading of homogeneously dispersed particles in a liquid carrier to ensure suitable moulding characteristics.[4,5,6] The appropriate addition of surfactants is, therefore, one of the most important requirements that significantly enhances the dispersion of the powder in the medium, which in turn results in an enhanced powder loading at a reasonably low viscosity. Fatty acids are the most widely used surfactants; they are introduced to disperse suspensions of alumina and paraffin wax.[7] Stearic acid is considered to be one of the most appropriate candidates due to its carboxylic functional-group anchoring at the powder surface and a carbon chain containing 18 carbon atoms dissolving into the binder matrix.[8] However, Johnson et al.[9] and Liu[10] have indicated that the use of small molecules such as stearic acid (the length of stearic acid is ~ 2.4 nm) does not provide true stabilization. According to this work, stearic acid is effective only in reducing the van der Waals attraction and/or the short-range forces, but it is not capable of creating a repulsive potential between the particles, and the resultant interparticle potential is attractive and the suspensions are weakly flocculated.

---

[*] Corresponding author. E-mail: ales.dakskobler@ijs.si

When the overall interaction potential in the suspension is attractive, there is a certain volume fraction of solids above which a continuous network is formed and a certain shear stress is needed to overcome the interparticle forces in order to induce flow. This shear stress or yield stress has been shown by many studies to be dependent on the interparticle forces, the volume fraction of solid, the particle size, and the particle size distribution[10,11,12]. Furthermore, all these parameters also affect the flow behaviour of these suspensions after the yield stress of the suspensions is exceeded. The yield stress also plays an important role in the mould-cavity filling step. The flow behaviour of these suspensions as determined with flow-curve measurements can usually be described as pseudoplastic flow.[13]

In the present study the effect of the particle size of the powders and the solids loading on the rheological properties of paraffin-wax suspensions are presented. As a model system we used three alumina powders with different granulations, and a paraffin wax, which has a liquid/solid phase transition at around 60℃. This wax was used for the preparation of suspensions that contained from 50 to 60 vol.% of powder.

EXPERIMENTAL WORK

Three alumina powders were used in this work: a submicron-sized alumina powder A16 SG (Alcoa, USA) and two micron-sized alumina powders CT 1200 SG and CT 800 (Martinswerk, Germany). The particle size for these three powders is given in Table 1.

Table 1: Particle sizes of the aluminas used for the preparation of suspensions

	Particle size (μm)		
	d_{10}	d_{50}	d_{90}
A16 SG	0.34	0.72	2.01
CT1200 SG	0.75	1.19	2.40
CT800 SG	1.27	3.28	6.50

The powder suspensions were prepared using INA 58/62 paraffin wax (INA, Croatia) with a melting point in the temperature range 60±2℃ as a liquid medium and stearic acid (Carlo Erba, Italy) as a surface-active agent. In the preparation of the slurry the alumina powder was dried at 140℃ for 4 h before compounding with a molten mixture of paraffin wax and stearic acid. After compounding, the suspensions were homogenized at 80℃ using a water-heated three-roller mill (Exact, Germany). Three passes through the gap between the rollers was sufficient for the preparation of homogeneous suspensions, as was experimentally determined in our laboratory.[14]

For each alumina powder the amount of stearic acid needed for the preparation of the suspensions was calculated per surface area of the particulate alumina powder. A total of 0.5 mg of stearic acid per square meter of the surface of the powder was used for the preparation of the suspensions[15]. The volume fractions of solids in the suspensions were 50, 55 and 60 vol.%, respectively.

After the preparation of the paraffin-wax suspensions the rheological properties were measured. The rheological properties were analysed with a MCR 301 rheometer (Anton Paar, Austria) using a cone-and-plate geometry and a PK-25-1 sensor system at 70℃. In order to determine the flow curves of these suspensions, a shear-rate range from 0.1 to 100 s^{-1} was selected. To determine the yield stress of the paraffin-wax suspensions, controlled shear-stress measurements were performed in the shear-stress range from 1 to 30000 Pa.

RESULTS AND DISCUSSION

In Fig. 1 the flow curve for the 60 vol.% A16 alumina, paraffin-wax suspension is presented. The suspension exhibits pseudoplastic behaviour, characteristic for such suspensions. The viscosity of this suspension is 5.7 Pas at a shear rate of 100 s^{-1}. As pointed out in the introduction, these suspensions are weakly flocculated due to the insufficient thickness of the stearic barrier provided by the adsorbed stearic acid. Consequently, at rest the particles lie in a secondary minimum because of the attractive interparticle interaction. To overcome these attractive interactions a certain shear stress – yield stress is needed to overcome the interparticle forces in order to induce flow. The measured yield stress for the 60 vol.% A16 alumina, paraffin-wax suspension was 185 Pa (Fig. 2). According to our experience with the preparation and use of such suspensions, the presented rheological properties of the suspension make possible the successful moulding of specimens. However, these rheological properties are the upper limit for successful shaping using LPIM.

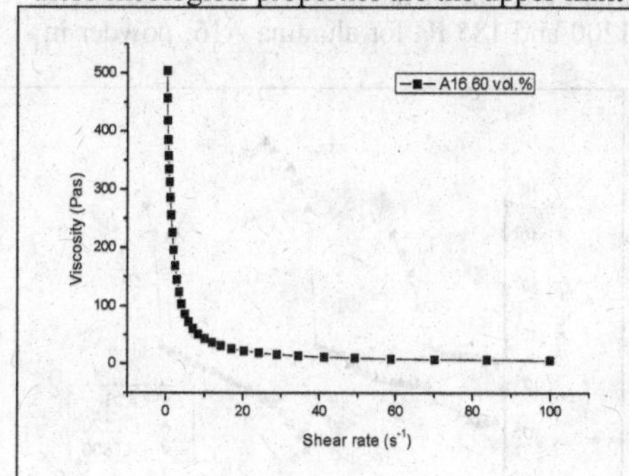

Fig. 1: Flow curve for 60 vol.% alumina, paraffin-wax suspension at 70℃.

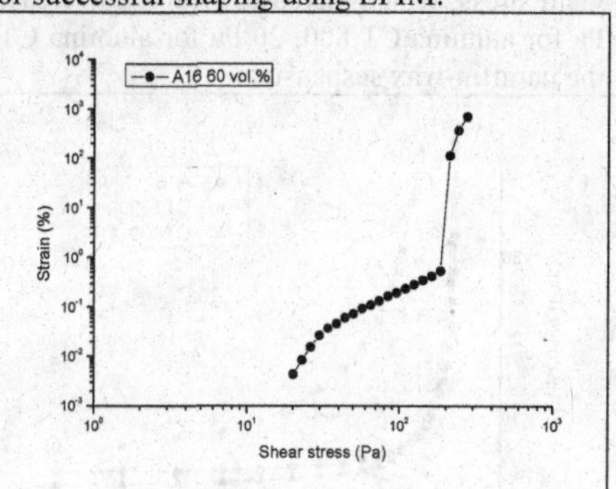

Fig. 2: Yield stress measurement of 60 vol.% alumina, paraffin-wax suspension at 70℃.

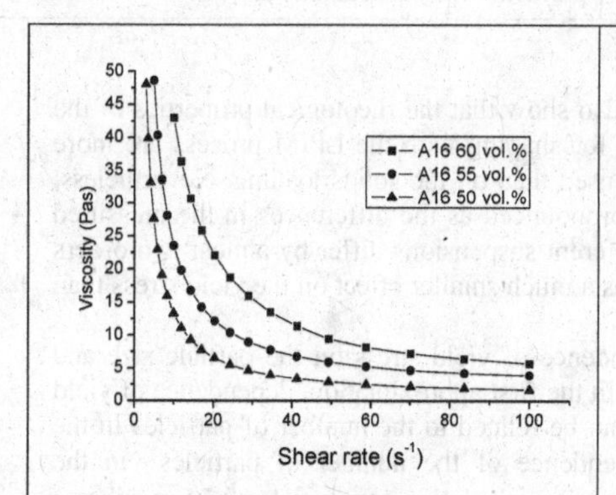

Fig. 3: Dependence of flow-curves on the solids loading of the prepared A16 alumina, paraffin-wax suspensions at 70℃.

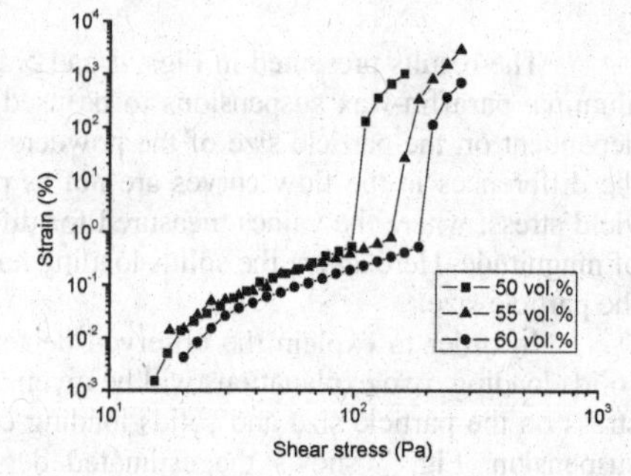

Fig. 4.: Dependence of the yield stress on the solids loading of the prepared A16 alumina, paraffin-wax suspensions at 70℃.

The rheological properties of alumina, paraffin-wax suspensions can be tailored either by changing the solids loading or by using aluminas with different particle sizes. In Fig. 3 the flow curves are presented, and in Fig. 4 are the yield-stress measurements for A16 alumina suspensions with 50, 55 and 60 vol.% of solids loading. With an increased solids loading the viscosity is increased across the whole range of shear rates (Fig. 3). This is also true for the yield stress, which is increased with a higher solids loading, and is 98 Pa for 50 vol.%, 142 Pa for 55 vol.% and 185 Pa for 60 vol.% of A16 alumina powder in the paraffin-wax suspension.

The influence of particle size on the flow curves and yield stress of alumina – paraffin-wax suspensions containing 60 vol.% of alumina A16 (d_{50}= 0.7 μm), alumina CT 1200 (d_{50}= 1.2 μm) and alumina CT 800(d_{50}= 3.2 μm), are presented in Fig. 4 and Fig. 5. Fig. 4 presents the flow curves, from which it can be concluded that with a decreased particle size the viscosity is higher across the whole shear-rate range investigated. This is also true for the yield stress, which increases with the decreased particle size of the powders used, and is 3.2 Pa for alumina CT 800, 20 Pa for alumina CT 1200 and 185 Pa for alumina A16 powder in the paraffin-wax suspension.

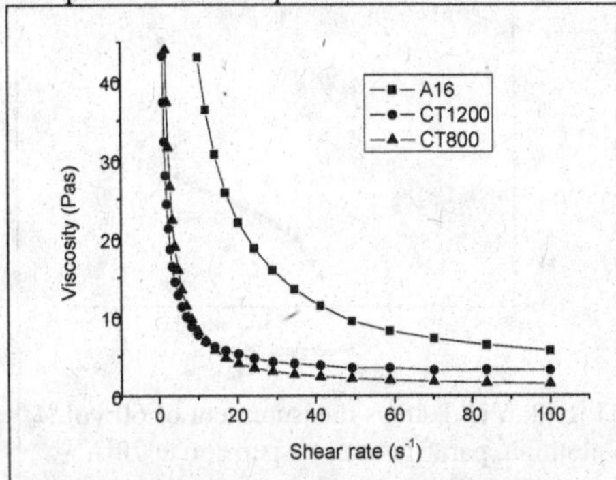

Fig. 5: Dependence of flow-curves on the particle size of the prepared 60 vol.% alumina, paraffin-wax suspensions at 70°C.

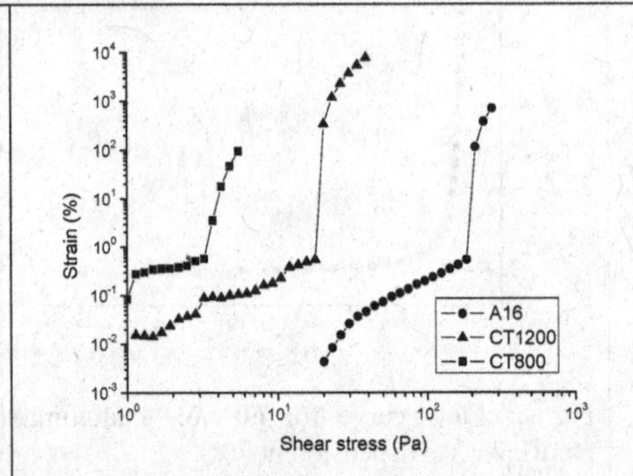

Fig. 6: Dependence of the yield stress on the particle size of the prepared 60 vol.% alumina, paraffin-wax suspensions at 70°C.

The results presented in Figs. 4 and 5 also show that the rheological properties of the alumina, paraffin-wax suspensions to be used for shaping with the LPIM process are more dependent on the particle size of the powders used than on the solids loading. Nevertheless, the differences in the flow curves are not as pronounced as the differences in the measured yield stress, where the values measured for different suspensions differ by almost two orders of magnitude. Here again, the solids loading has a much smaller effect on the yield stress than the particle size.

In order to explain the observed dependence of yield stress on the particle size and solids loading, two explanations will be given. In the first approximation, dependence of yield stress on the particle size and solids loading can be related to the number of particles in the suspension. Fig. 7 shows the estimated dependence of the number of particles in the suspension volume unit on solids loading (for easier calculations as particles 0.7 μm spheres were used). Number of particles in volume unit of the suspension is linearly increasing with increased solids loading. The change in the number of particles in the range of prepared suspensions solids loading from 50 (2700 particles) to 60 vol.% (3400 particles) is in the same range as the change in the yield stress measured for 50 and 60 vol.%, which were 98 Pa and 185 Pa, respectively. In both cases the change was less than 100%.

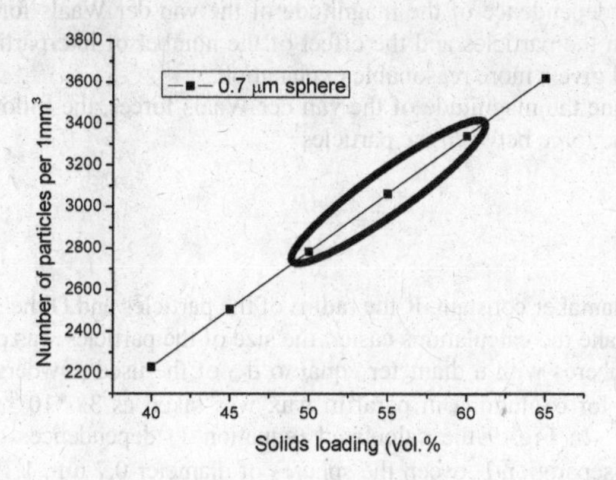

Fig. 7: Dependence of the number of particles per 1 mm^3 on the solids loading of the suspension. (Number of spheres with a diameter of 0.7 mm per 1 mm^3. Range of interest is marked with ellipse.)

In a similar way, the effect of the particle size on the yield stress was illustrated. The number of particles was calculated as a number of spheres with the diameter from 0.1 to 5 μm, which have equal volume to the volume of 5 μm sphere. In the range of interest, the yield stress changed from 3.2 Pa for 60 vol.% alumina CT800 – paraffin-wax suspension to 185 Pa for 60 vol.% alumina A16 – paraffin-wax suspension (see Fig. 6) and the number of particles changed from 3.8 for 3.28 μm sized sphere to 364 for 0.7 μm sized sphere. In both cases the change by two orders of magnitude was determined.

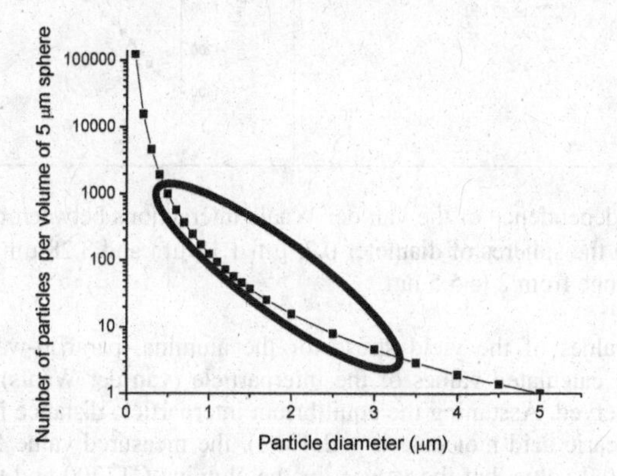

Fig. 8: Dependence of the number of particles with diameter from 0.1 to 5 μm occupying the volume of a 5-μm sphere. (Range of interest is marked with ellipse.)

In spite of the relatively good agreement for the presented explanation, it is necessary to consider the yield-stress dependence from the point of view of the attractive interparticle

interactions, which are the reason for the formation of the attractive interparticle network in the suspensions. The yield stress is a measure of the strength of the formed interparticle network, which is directly related to the van der Waals forces acting between the particles. Understanding the dependence of the magnitude of the van der Waals forces on the particle size acting between the particles and the effect of the number of interparticle interactions on the yield stress will give a more reasonable explanation.
In order to determine the magnitude of the van der Waals forces, the following equation was used to calculate the force between the particles[16]:

$$F = \frac{AR}{12D^2} \tag{1}$$

with A being the Hamaker constant, R the radius of the particles and D the separation between the particles. To make the calculations easier, the size of the particles was chosen such that all the particles are spheres with a diameter equal to d_{50} of the used powders. The value of the Hamaker constant for α-alumina in paraffin wax was taken as $3.6*10^{-20}$ J, as proposed by Boergstrm et al [17]. In Fig. 9 the calculated (equation 1) dependences of the interparticle interactions of the separation between the spheres of diameter 0.7 μm, 1.19 μm and 3.28 μm are presented.

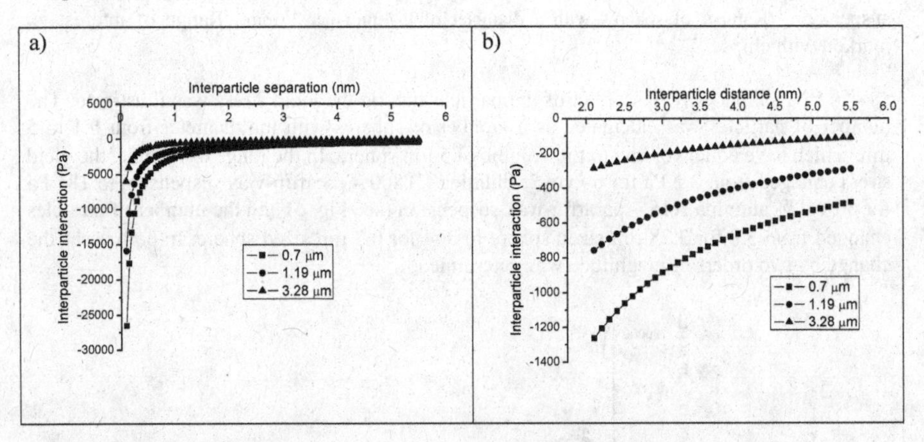

Fig. 9: Calculated dependence of the van der Waals interactions between the particles for the separation between the spheres of diameter 0.7, μm 1.19 μm and 3.28 μm: a) range from 0.1 to 5.5 nm and b) range from 2 to 5.5 nm.

If the measured values of the yield stress for the alumina, paraffin-wax suspensions are compared with the calculated values of the interparticle (van der Waals) interactions, large differences are observed. Assuming the equilibrium interparticle distance in the suspension is 4 nm (length of stearic acid molecule is ~ 2.4 nm), the measured value for alumina A16 is 0.28 of the calculated value, but the values for the alumina CT1200 and the alumina CT800 are only 0.05 and 0.02 of the calculated values, respectively. These results are shown in Fig. 10, where the drastic change of the ratio of the measured/calculated values (RMC) is shown. Suspensions exhibiting low values of the presented ratio enable the preparation of suspensions with an almost predicted maximum solids loading for monodispersed powders[18], with properties suitable for shaping with the LPIM process. Contrary, A16 alumina suspensions with suitable properties for LPIM can be prepared with up to 60 vol.% of A16 alumina, which is considerably lower than is the predicted maximum solids packing for monodispersed powders.

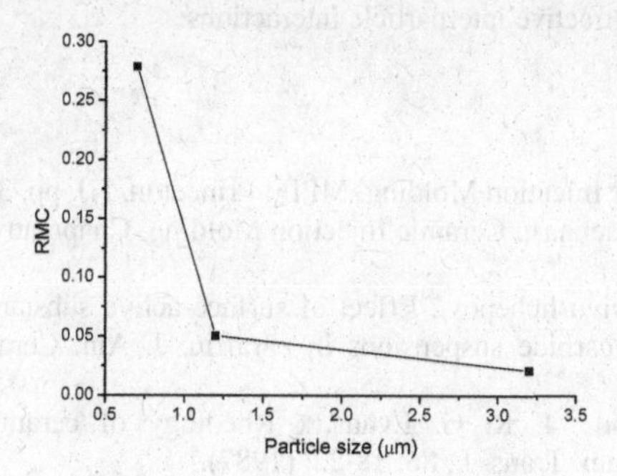

Fig. 10: Dependence of the ratio between the measured yield stress and calculated interparticle interaction (RMC) on the particle size.

Furthermore, these results showed that for the preparation of the alumina – paraffin-wax suspensions, the attractive interparticle interactions must be taken into account. For large particles the attractive interparticle interactions can be neglected, while for smaller particles (<1 μm) these interactions determine the rheological properties, and with decreasing particle size they become more important. This results in a lowering of the solids loading of the suspensions with a decreased size of the particles used. The major reason for decreased solids loadings are the attractive interparticle interactions between the sub-micrometer particles in the alumina – paraffin-wax suspensions.

The 60 vol.% alumina A16 – paraffin-wax suspension can be successfully shaped by LPIM. According to the results presented, the suspensions prepared with a size of alumina particles smaller than alumina A16 must contain a lower solids loading than A16 alumina suspensions to make the shaping by LPIM possible. This is mainly due to the more pronounced attractive interparticle interactions in the suspension, while the contribution of the effective particle volume (Φ_{eff}) on the solids loading in the observed particles size range from 0.3 to 4 μm is very small.

CONCLUSIONS

The flow-curve measurements of the suspensions containing alumina particles in molten paraffin-wax reveal the pseudoplastic behaviour, which is more pronounced for the suspensions prepared with a smaller particle size and a higher solids loading.

The yield stress of the alumina, paraffin-wax suspensions was measured. The origin of this yield stress is the "small" attractive interaction between the particles and the formation of the attractive particle network in the suspension at rest. The yield stress of the suspensions increased with a decrease in the particle size, and an increased solids content used for the preparation of the paraffin-wax suspensions.

The only parameter that was changed during this study was the number of particles per volume unit and, consequently, the number of attractive interparticle interactions in the suspensions per unit volume. The results showed that during the preparation of the alumina, paraffin-wax suspensions the attractive interparticle interactions must be taken into account.

For large particles the attractive interparticle interactions can be easily neglected, but for smaller particles (<1 μm) these interactions determine the rheological properties, and with decreasing particle size they become more important. With a decreasing size of the particles used the solids loading of the suspensions is lowered. The major reason for the decreased solids loadings are the attractive interparticle interactions.

REFERENCES

[1] R. M. German, Powder Injection Molding, MPIF, Princeton, NJ, pp. 3-22, (1990).
[2] B. Mutsoddy and C. Beebhas., Ceramic Injection Molding, Chapman & Hall, London, pp. 1-9, (1992).
[3] R. Lenk and A.Ph. Krivoshchepov., Effect of surface-active substances on the rheological properties of silicon carbide suspensions in paraffin. J. Am. Ceram. Soc., **83**, 273-276, (2000).
[4] M. J. Ediringshe and J. R. G. Evans, , Rheology of ceramic injection moulding formulations. Br. Ceram. Trans. J., **86**, 18-22, (1987).
[5] M. J. Ediringshe , H. M. Shaw and K. L. Tomkins, Flow behaviour of ceramic injection moulding suspensions. Ceramics Int., **18**, 193-200, (1992).
[6] V. M. Moloney, D. Parris and M. J. Edirisinghe, Rheology of zirconia suspensions in non-polar organic medium. J. Am. Ceram. Soc., **78**, 3225-3232, (1995).
[7] T. Y. Chan and S. T. Lin, Effects of stearic acid on the injection moulding of alumina. J. Am. Ceram. Soc., 78, 2746-2752, (1995).
[8] S. Novak, K. Vidović, M. Sajko and T. Kosmač, Surface modification of alumina powder for LPIM. J. Eur. Ceram. Soc., **17**, 217-223, (1997).
[9] R. E. Johnson Jr. and W.H. Mossison Jr., Ceramic powder dispersions in non-aqueous systems. Adv. Ceram., **21**, 323-348, (1987).
[10] D. Liu, Effect of dispersants on the rheological behavior of zirconia-wax suspensions. J. Am. Ceram. Soc., **82**, 1162-1168, (1999).
[11] R. Buscall, I. J. McGowan, P. D. A. Mills, R. F. Stewart, D. Sutton, L. R. White, and G. E. Yates, The rheology of strongly flocculated suspensions. J. Non-Newtonian Fluid Mech., **24**, 183-202, (1987).
[12] Z. Zhou, M. J. Solomon, P. Scales and D. V. Boger, The Yield stress of concentrated flocculated suspensions of size distributed particles. J. Rheol., **43**, 651-671, (1999).
[13] R. M. German, Powder Injection Molding, MPIF, Princeton, NJ, pp. 147-172, (1990).
[14] A. Dakskobler, Karakterizacija suspenzij za nizkotlačno injekcijsko brizganje keramike. B. Sc. University of Ljubljana, (1995).
[15] L. Gorjan, Priprava suspenzij za izdelavo korundnih izdelkov z brizganjem. B.Sc. University of Ljubljana, (2006).
[16] J. N. Israelachvili, Intermolecular and surface forces, Academic Press, London, p. 139, (1985).
[17] L. Bergstrom, A. Meurk, H. Arwin, and D. J. Rowcliffe, Estmation of Hamaker constants of Ceramic Materials from Optical Data using Lifsithz theory, J. Am. Ceram. Soc., **79**, 339-348, (1996).
[18] R. M. German, Powder Injection Molding, MPIF, Princeton, NJ, pp. 102-136, (1990).

PREPARATION OF HIGHLY CONCENTRATED NANOSIZED ALUMINA SUSPENSIONS FOR SPRAY-DRYING

S. Cottrino, J. Adrien and Y. Jorand
MATEIS Laboratory of INSA Lyon
Villeurbanne, France

ABSTRACT

As it is of major importance to use highly concentrated slurries in spray-drying processes, procedures to obtain nano-alumina suspensions have been investigated : determination of suspending medium pH, dispersant characteristics and ratio, and optimization of dispersion procedure.

The rheological behavior of the slurries was characterized following various dispersant additions and ultrasonic treatments. Spray-drying was conducted in a 20 kHz ultrasonic atomizer giving a typical granule size of 80 μm. The resulting granules were evaluated in terms of flowability, morphology and microstructure using classical normalized tests, mercury intrusion porosimetry and tomography observations. Slurry viscosities were found to fit the Casson law and can be described using two parameters, the yield stress (τ_o) and viscosity limit (η_∞). Nanosized suspension rheology was dependant on the dispersant molecular weight. After ultrasonic treatments, a quasi-Newtonian behavior was observed with low molecular weight dispersants. Finally, suspension with volume fraction as high as 45 vol. % were obtained. The spray drying experiments showed that it is possible to obtain near solid granule morphology with well dispersed suspensions, i.e. with a quasi Newtonian behavior, but combined with high solids contents.

This work confirmed that the average molecular weight of the dispersant used in nanosized particle dispersion is of primary importance. Moreover, it has been shown that the efficiency of ultrasonic de-agglomeration treatments is strongly dependent on the average molecular weight of the dispersant.

INTRODUCTION

Nanostructured materials are likely to provide enhanced properties, especially from a mechanical point of view. However, the major difficulty in ceramic fabrication is to produce fully dense parts while maintaining nanoscaled grains. Dry pressing is a widely used shaping method for ceramic parts, the advantages being a low production cost and a high productivity. The main disadvantage is that it is more prone to induce large defects in green parts than wet forming techniques[1,2]. This is an important parameter in the production of structural parts, as it is well established that sintering of fully dense bodies requires a grain size of the same order of scale as that of the largest porosities[3]. Hence, the use of nanoparticles is beneficial only if one can produce green bodies with a population of porosities as close as possible to the initial grain size. Given the present state-of-the-art, it is obvious that this objective is still far from being achievable even for microparticles.

As free flowing powders are less subjected to packing defect development, fine powders granulated by spray-drying is an established granulation technique in the ceramic industry. For this reason, promotion of the use of nanopowders in ceramic requires the development of specific spray-drying methods. In order to avoid the presence of large pores in compact microstructure, it is of benefit to consider which granule characteristics are optimum. The ideal granule should be spherical with a diameter of about 80-100 μm to ensure high flowability[4]. It should also have a solid morphology and a mechanical behavior such that the inter-granule porosities are resorbed. The two former stages have been extensively studied by Walker et al.[5,6] for a micronic alumina powder. Spray-drying of low yield stress slurries, i.e. those which are well deflocculated, produces hollow granules (figure 1). During the first stage of droplet drying, an outer dry rigid shell is formed. Then, as final drying proceeds, no further shrinkage is possible and particles thicken the granule wall leaving an internal void. The major conclusion of their study is that a de-optimization of the slurries is needed. Hence, the high slurry yield stress hinders particles dragging by water during drying and solid granules are obtained. This corresponds to a transition from a slightly flocculated state, with small flocs units, to a continuous floc network resulting from the increase of solid content. The use of de-optimized slurries avoids the occurrence of a range of defects due to the resulting large inner granule voids. However, a reduction of the primary particle compacity is obtained, i.e. the inter-particle porosities are larger than the primary particles diameters. As mentioned previously in this introduction such green microstructure is not adapted to fine microstructure development.

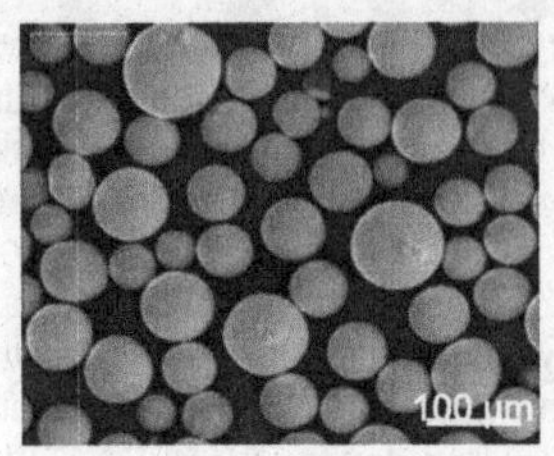

Figure 1 : Typical hollow granule

The objective of this paper is to investigate formulations of alumina nano-powder slurry that could produce solid granules and a high primary particle packing densities. The preferred solution to reach this objective is very high solid content deflocculated suspensions route. Hence the main results for high solid ratio slurries realization, are presented in the following.

The dispersion viscosity increases dramatically with solid loading as it approaches a maximum at which flow is blocked. This volume fraction limit depends on the dispersion system considered[7]. Many parameters may affect this value, but for a given system the dispersion state is of a major importance. A well dispersed system can yield a loading limit as high as 60 vol % compared to approximately a few ten percents for the same coagulated system. On the other hand, it is known that stable nano-powder dispersions are much more difficult to obtain than micronic ones : for a same loading ratio the separation distance between particles decreases with the particle diameter[8]. Hence, nano-dispersions require the combination of "electrostatic" and "steric" repulsion mechanisms.

The apparent charge density on the surface of oxide particles varies with the pH. At low pH, positive sites are prominent whereas at high pH surfaces are negatively charged. At the zero point charge (zpc), the surface is neutral, the number of positive sites equals the number of negative sites. For gamma alumina, the zpc has been measured near pH 9.8[9].

PAA and PMAA are widely used for alumina slurry stabilization. Recent studies[9,11] have shown that these polyelectrolytes, which combine the two dispersion mechanisms, are efficient to stabilize alumina nanopowders. The following brief discussion pertains to the way of adsorption of polyelectrolytes.

The structure of polyelectrolytes has many carboxylic acid sites (functional groups) and the fraction of functional groups dissociated varies with solvent conditions (pH, ionic strength...). The dissociation and the negative charge characteristics of the polymer increase with pH. For PAA in water, a nearly full dissociation is reached at pH 11 [10]. On the other hand, it has been shown that adsorption ratio of PAA on alumina surfaces (either alpha or gamma forms) at pH 3 is approximately twice as large as that at pH 11[9,10,11,12]. This leads to a "pancake" type adsorbed layer at low pH and a "brush-like" conformation at high pH. The former configuration is reported to have the strongest steric dispersion effect [Bowen]. Similar observations were reported by Cesarano et al.[12] with PMAA in alpha alumina-water system. These results lead to the conclusion that in order to stabilize alumina dispersions, a compromise is necessary between the polymer adsorption concentration and the conformation on the nano-particle surface. In conclusion, the most efficient pH dispersion range is between 6 to 9[9,11].

In nano-suspensions a major limitation of maximum solids content arises from the excluded volume that is formed around particles by the polymeric dispersant. This questions on the most suited chain length for steric dispersants. A part of the answer is given by the study of Studart et al. as they shown that the optimum for this adlayer thickness is close to 4 nm in nanoalumina suspensions[13]. It can be concluded that low molecular weight constitutes the best choice if brush-like conformation are considered. Indeed, K. Lu and co-author obtained solid loadings as high as 45 vol. % with nanoalumina suspensions using PAA 1800. Nevertheless, in the case of pancake conformations, higher Mw are likely to provide good results.

De-flocculation treatments are of first importance in slurry preparation, since they reduce or suppress the flocs which are initially present, even if the system is optimized to ensure dispersion stability. The flocs consist in very porous entities with entrapped water that can not be mobilized during shearing. Their suppression permits a substantial increase of the solid content for a given viscosity. Ultrasonic treatment is known to be an efficient method for suspension de-flocculation, but the effects of this technique in ceramic processing are somewhat not well defined. Many authors have reported a benefit of sonification even for very short durations and for various ceramic dispersion systems: micrometric alumina[14], nanometric zirconia[15], SiC[16], and TiO2[17]. Among these investigations, it has been observed that either high power input or long treatment time induce slurry re-flocculation[14,15,16]. In these papers, the preferred explanation for this phenomenon is that the dispersant chains are damaged by the ultrasound application, which means that this detrimental effect is more pronounced when the chains are long. Nevertheless, the combined influence of PAA Mw and sonification has been studied by Sato et al. and it appears that sonificated suspensions provide better results, in terms of viscosity and agglomerate size, with PAA molecular length as high as Mw 8000 compared to the 2000 usually employed.

In order to achieve a high loading ratio, we have first focused our study on the dispersion condition optimization: pH, dispersant type (PAA or PMAA, chains length) and dispersant amount. Experiments were then conducted in order to clarify ultrasound effects on polyelectrolyte deflocculated nano-dispersion, especially in the case of high molecular weight dispersants.

EXPERIMENTAL PROCEDURE

Starting materials

The powder used in this study was a gamma alumina nanopowder (Nanotek) with a specific surface area of around 35 m²/g and an average particle size of 47 nm. This nanopowder is constituted of spherical particles which are favorable in obtaining low viscosity slurries (figure 2). It can also be shown however that it is largely agglomerated.

Polyacrylic acid (PAA) with average molecular weight (M_w) of 2,000 (Reactolab), 5,000 (Reactolab), 100,000 (Sigma-Aldrich) and polymethacrylic acid (PMAA, Darvan C-N) with mol. wt. 15,000 were used as polymer dispersants.

Figure 2: Scanning electronic micrograph of Nanotek alumina.

Preparation of suspensions

The dispersant (PAA or PMAA) was mixed with water and after 3 min of homogenization, the pH was adjusted with NH_4OH to the desired level. The powder was then added to this suspending medium and stirred for 30 min, since Chen et al. have shown that adsorption process on alumina surfaces is saturated after 20 min[10]. Finally the standard protocol includes an ultrasonically irradiation of 10 min with a 100 W output power and 80 % pulsed mode. Ultrasonification was conducted in an ultrasonic flow reactor (Sonitube, Sodeva) operating at 35 kHz with a flow rate of 16 liter per hour. This apparatus has two main advantages, it allows a continuous process and the acoustic field is rather uniform for a given cross section of the treatment tube. In comparison to an immersed probe, a more homogeneous treatment is expected. Rheological characterization of the slurries was carried out with a co-axial rheometer (Haake VT 510). For each suspension, the yield stress extrapolated on the basis of a Casson model and the apparent viscosity at shear rate of 100 s^{-1} were selected. To avoid any error linked to the slurry's treatments, solid loading was then measured with a halogen moisture analyzer (Mettler Toledo HR83).

Spray drying and granules characterization

Slurries were granulated with a 20 kHz ultrasonic spray-dryer, with inlet and outlet temperatures maintained respectively at 160°C ± 5°C and 80°C. This kind of atomizer presents the advantage to produce low velocity spray, of around 0.2–0.4 m/s in comparison with the 10 to 20 m/s droplet velocity of conventional two fluid nozzles. Long residence times and representative granule sizes of 80 μm are therefore achievable even with a laboratory spray-drier. Moreover, droplets with high sphericity and uniform size distribution are achieved[18].

Each atomization batch was characterized by measuring granule size distribution with a laser granulometer equipped with a dry powder feeder (Malvern, Mastern sizer 2000, Sirocco module). Flowability, loose- and tapped-bulk density were obtained using ISO Standard protocols.

Granule bulk density and granule inter-particle pore size distribution were obtained by mercury intrusion porosimetry (Micromeritics autopore II 9400). Pore diameter D_p was determined from the instrument pressure P using the Washburn equation:

$$D_p = \frac{-4\gamma\cos\theta}{P}$$

Using γ = 485 mN/m for the surface tension of mercury and a wetting angle θ = 140°. Figure 3 shows a typical bimodal pore size distribution for granulated materials. The intergranule porosity, due to interstitial voids resulting from the packing of the granules, was not considered : it depends not only on granule characteristics but also on cell filling conditions . The smaller pores range are interstitial voids resulting from the packing of primary particles within granules.

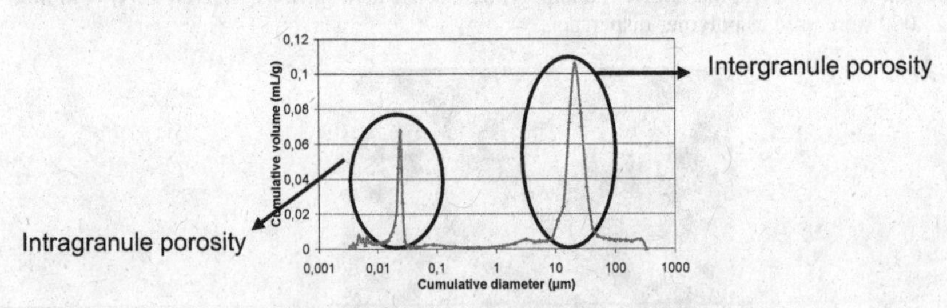

Figure 3 : Typical mercury porosity curve

Finally, in order to estimate the size of inner granule void and the percentage of these voids compared to the apparent granule volume, X-ray computed tomography was used. This technique is an alternative way to determine porosity, pore size and visualize the microstructure of an entire granule with details comparable to the resolution of optical microscopy instruments in a non destructive way. As a result, individual pores are visualized and the true internal morphology of granules is revealed. X-ray tomography was performed using a v/tome/x X-ray microtomograph. This commercial device includes a nanofocus transmission X-ray tube (W target). A small sample of each granule batch was scanned with a resolution of 3μm, with radiographs taken over 360°. The tomograph was operated at 80 keV and 160 μA and typical acquisition time was about 30min. 3-D images reconstruction and quantitative measurements were conducted using respectively "Datos_rec" (Phoenix) and Image J (Wayne Rasband) softwares.

Prior to all the analyses, granules were conditioned in a dessiccator for at least one week at a constant humidity of 5 %.

RESULTS AND DISCUSSION

Optimum pH determination

The behavior of Nanotek alumina powder without dispersant, was studied from pH 5.5 to 11.5 (figure 4). At pH 9, the viscosity reaches a maximum, corresponding to slurry flocculation. This result is in agreement with zeta potential data obtained by Azar[20] et al. with Nanotek alumina, the isoelectric point is reported at pH 9.

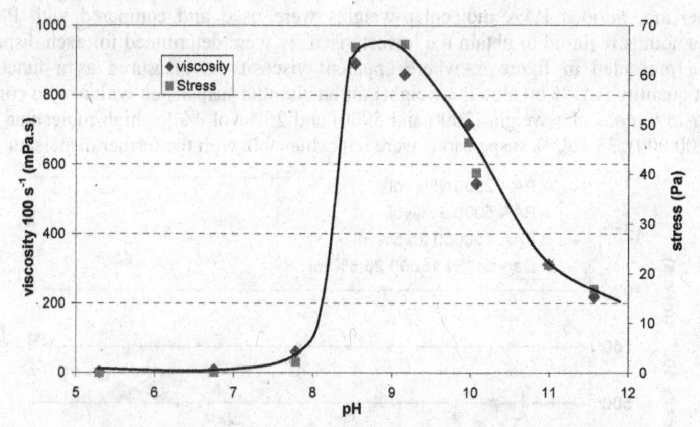

Figure 4: Behavior of powder without dispersant (12.5%vol)

On the basis of these results it can be concluded that the suspension pH must be lower than 7. Nevertheless the effects of dispersant additions must be considered : on the one hand, either PAA and PMAA additions shift the zeta potential vs. pH curve towards low pH values; on the other hand, as previously discussed, high pH values are beneficial in obtaining a high dissociation ratio and a favorable "brush-like" conformation. Bowen et al.[9] have investigated dispersion of gamma alumina with PAA additions up to pH 10. For pH 10, they reported an evolution of the zeta potential from -5 mV to about – 40 mV for additions of a few percent of PAA 2000 addition. As a consequence we have considered the dispersion behavior of Nanotek alumina with PAA 2000 and Darvan C-N for higher pH values. For comparison, similar experiments were conducted at pH 9, which is close to the value usually retained for such systems[9,11,12]. These results presented in fig. 5, show that lower viscosities are obtained with PAA 2000 and pH has only a little influence on its behavior. In contrast, Darvan C-N is more sensitive to pH range, and as pointed out by Cesarano et. al the best results are obtained at pH 9. In the following experiments the dispersion pH will be kept at a value of 9.

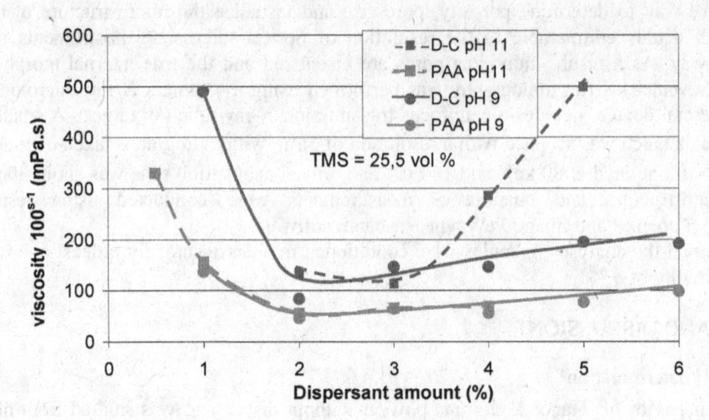

Figure 5: Dependence of viscosity on dispersant amount for pH 9 and 11

Effect of dispersant molecular weight

In order to determine the effect of dispersant chain length on the rheological properties of nano-dispersion, various PAA molecular weights were used and compared with PMAA. The optimum amounts required to obtain the lowest viscosity were determined for each dispersant. The results are presented in figure 6, where apparent viscosity is measured as a function of the dispersant quantity (wt. % on alumina weight) for an alumina suspension with a solid content of 35 vol. % for low molecular weight (2000 and 5000) and 25.5 vol. % for high molecular weight (15 000 and 100 000). 35 vol. % suspensions were not achievable with the former dispersant.

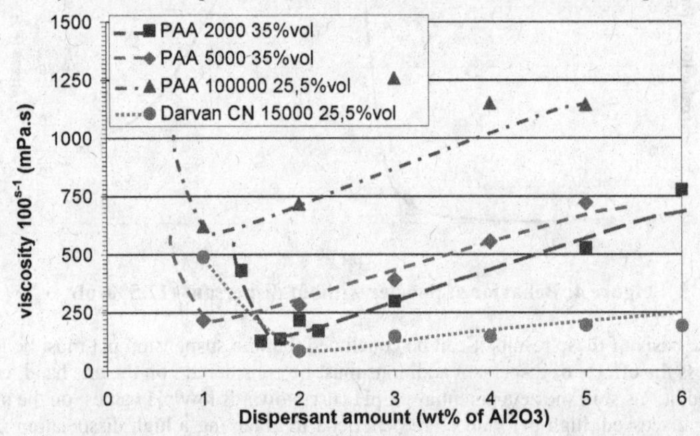

Figure 6: Dependence of viscosity on dispersant amount for several dispersant Mw

A large increase in suspension viscosity with the dispersant molecular weight was observed. In fact, even with a decrease in solid content (35 to 25,5), a larger viscosity is obtained with large molecular weight. Little viscosity evolution was reported by Cesarano et al for a similar PAA Mw range. This can be ascribed to the much larger particle separation distance in a micro-dispersion in comparison to that in a nano-dispersion, which impedes particle bridging by the large polymer chains. The above results tend to indicate a reduction in the optimum dispersant amount with the dispersant molecular weight. Assuming the number of adsorption sites constant for a given dispersed powder, the weight ratio at the optimum for two polyelectrolytes with different Mw should be in the molecular weight ratio of the dispersant considered. This contrasts with our experimental results, which indicates that the molecular chain of the dispersant evolves from a "brush" type to a "pancake" conformation as the dispersant Mw increases. In Figure 7 the same data are plotted as a function of the mole number of added polymer, showing that the number of PAA 100,000 chain adsorbed is more than two orders of magnitude less than that for the PAA 2,000.

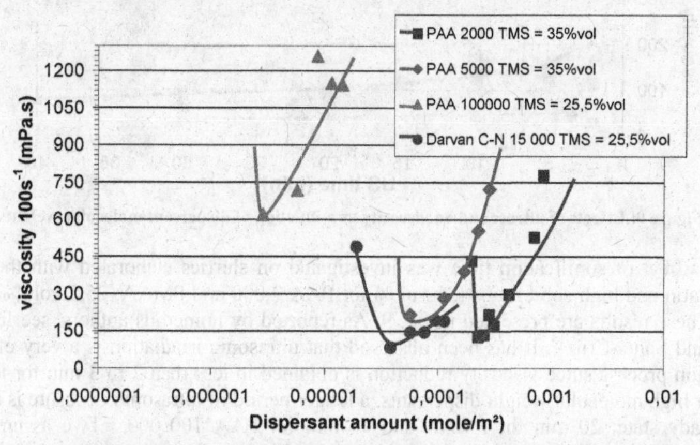

Figure 7: Dependence of viscosity on dispersant amount for several dispersant Mw

Effects of ultrasonic treatments

Figure 8 presents the effect of a 10 min ultrasonic treatment, labeled standard protocol in the experimental procedure section of this papaer, on the apparent viscosity for suspensions dispersed with PAA 2,000 or PMAA. Clearly, sonification reduces the agglomeration ratio and so the rheological behavior. No effect on the dispersant optimum value can be detected, indicating that sonification has no incidence on polymer adsorption.

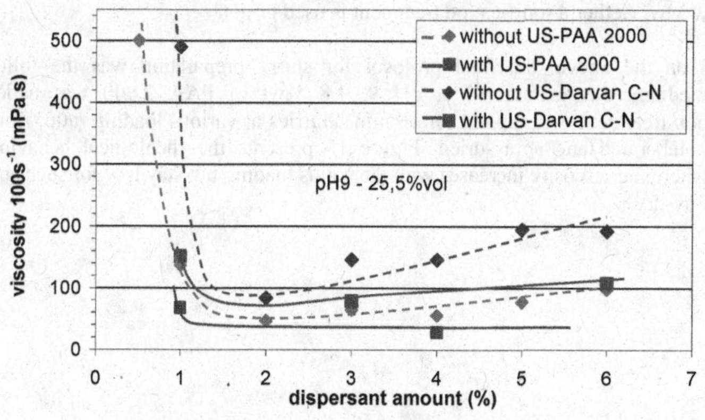

Figure 8: Effects of Ultrasound on dispersant behaviour

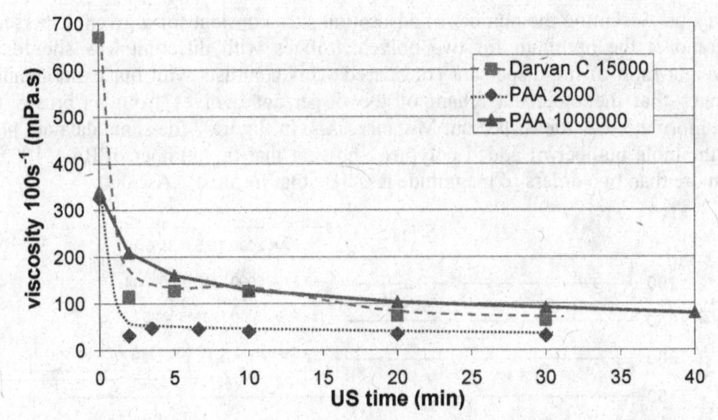

Figure 9: Effects of ultrasound on viscosity as a function of dispersant molecular weight

The effect of sonification time was investigated on slurries elaborated with the optimum dispersant ratio and high solid ratios (35 vol % for PAA 2,000 and PMAA; 25,5 vol. % with PAA 100,000). These results are presented in fig. 9. As reported by numerous authors, see for example Chartier[14] and Santa-Cruz[15], It has been observed that ultrasonic irradiation is a very efficient de-agglomeration process since viscosity reduction is obtained in less than 2 to 3 min for PAA 2,000 slurries. For high molecular weight dispersants, a longer period of ultrasonic exposure is required to reach a steady state: 20 min for PMAA and 40 min for PAA 100 000. Two assumptions are considered to explain this variation in de-agglomeration kinetics. First, as the dispersing medium is greatly affected by the addition of high molecular weight species, larger dissipation of the energy provided by bubble implosions generated by ultrasonic wave may be considered. As a consequence, the efficiency of the treatment is reduced and longer times are required. Secondly, as previously mentioned, large polymer chains may create bridging between particles and/or flocs which are difficult to break. This process, which proceeds probably by desorption, needs higher levels of energy, therefore longer treatment times are required. Sato et al.[17] have observed that the optimum in PAA Mw is located between 8,000 and 15,000 for deflocculation efficiency when combined with ultrasonic treatments of nano-TiO_2 dispersions. In contrast, nano-alumina dispersions offer best results for low Mw, wether a sonification treatment is used or not.

Based on the above, the final protocol for slurry preparation was the following: the suspending medium was maintained at pH 9, 1.8 %wt of PAA 2,000 was added and an ultrasonification treatment was applied for 10 min. Slurries at various loading ratio, from 28 to 45 vol % were elaborated and spray-dried. Figure 10 presents the rheological behavior of these suspensions where the viscosity increases with the solids loading but stay low for such solid content in order to spray-dry.

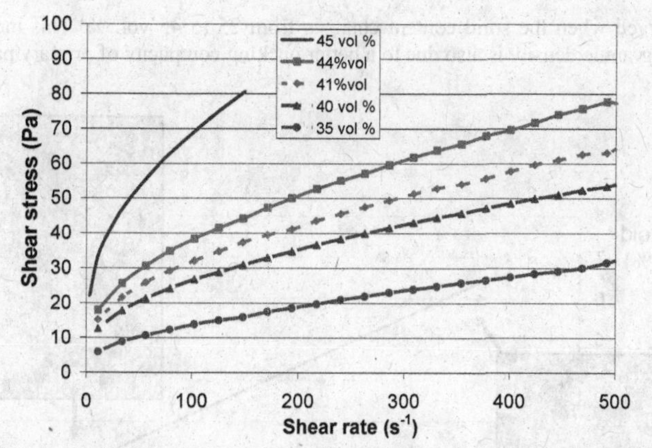

Figure 10: Rheological behavior of suspensions dispersed under the optimum conditions at various loading ratio

Spray-drying

In this part, the influence of solid content on the granules characteristics is studied. Figure 11 (graph a) shows the evolution of flowability, bulk tapped and loose density with feed material solid loading. These parameters increased as the solids concentration was increased.

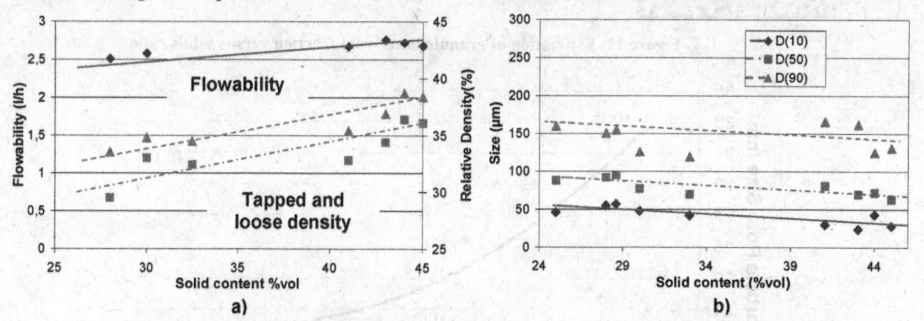

Figure 11: Evolution of flowability, bulk density and granule size with solids loading

In figure 11 (graph b), granule size distribution is also represented as a function of the solid content. A slight decrease in granule size distribution with the increase in solid loading is observed. This is a surprising result since it would be more natural that granule size increases with solid loading of the feed material[5]. This result may be due to the ultrasonic atomizing process, a slight dependence of the droplet size on slurry viscosity is commonly reported[18]. Therefore as slurry viscosity shows a large increase with solids loading, smaller droplets are produced. The fact that flowability and bulk density increase as the granule size decreases indicates that granule density has increased. This can proceeds from two causes, an increase of the packing compacity of primary particles and/or a reduction of the internal void.

The tomographic reconstructed slice presented in fig.12, demonstrates that the volumic void fraction is reduced as the slurry solid loading is increased. On the basis of a linear evolution of the internal void volume with solid loading, it can be estimated that a solid content of around 55 vol % is necessary to obtain solid granules[19]

Mercury intrusion porosimetry experiments reveal (see fig. 13) the dependance of interparticle pore size on slurry solid loading. A reduction in the pore size from 30 nm to less than

20 nm is observed when the solid content changes from 25 to 45 vol. %. This indicates that the increase of the granule density is also due to a better packing compacity of primary particles.

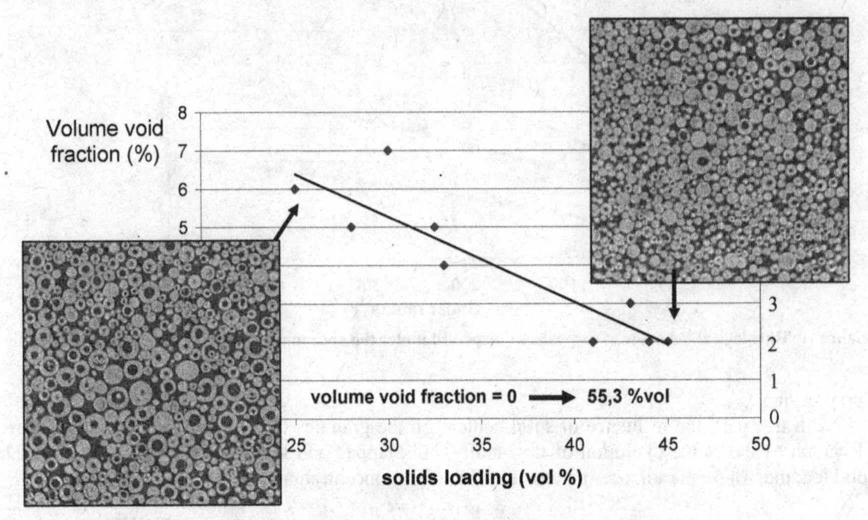

Figure 12: Estimation of granule inner void fraction versus solids ratio

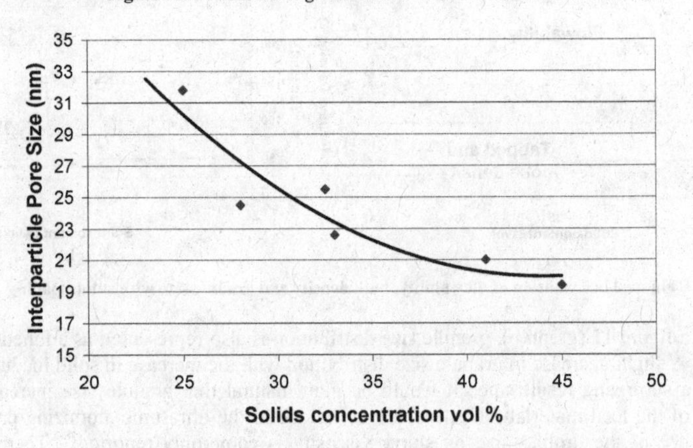

Figure 13: Evaluation of interparticle pore size versus solids ratio

CONCLUSIONS

Colloidal processing of nanosized alumina powders has been investigated in order to obtain spray-dried solid granules. It was found that the most desirable conditions to achieve 45 vol. % solid loading suspensions were identified to be pH 9 with 1.8 wt% of Al_2O_3 PAA 2,000 in combinaison with a 10 min sonification treatment. These suspensions were obtained with rheological characteristics favourable to spray-drying, as low as 500 mPa.s apparent viscosity at 100 s^{-1} and shear stress of 18 Pa.

This study confirms that sonification treatments are very efficient in dispersing nanometric suspensions. The viscosity of nano-alumina slurries following ultrasonic irradiation was significantly lowered, yield stress and apparent viscosity at 100 s^{-1} were reduced respectively of 77 % and 88 %. However, it seems that this treatment has no incidence on the value of the optimum dispersant ratio. High molecular weight dispersants are less efficient to improve rheological properties of nano-suspension in any case, with or without ultrasonic treatments. Moreover, the saturation of the viscosity reduction is obtained for longer ultrasonic treatments if high Mw molecular weight dispersants are employed.

Additionally, a study of the characteristics of spray-dried granules was also performed. It was shown that the increase of solids ratio slurries induces a large decrease in inner voids granules and interparticle porosity.

REFERENCES

[1] A. Broese Van Groenou, Compaction of ceramic powders, *Powder technology*, **28**, 221-228 (1981)

[2] L.T. Kuhn, F.F. Lange, A model for powder consolidation, *J. Am. Ceram. Soc.*, **74**, [3], 682-685 (1991)

[3] Kingery, Bowen, Uhlmann, Ed. John wiley & sons, Introduction to ceramics, p 488 (1976)

[4] "Particle Packing characteristics", Publisher : Metal Powder Industry, (1989)

[5] William J. Walker Jr., James S. Reed, Influence of slurry parameters on the characteristics of spray-dried granules, *J. Am. Ceram. Soc.*, **82**, [7], 1711-19 (1999)

[6] William J. Walker Jr., James S. Reed, Influence of granule character on strength and weibull modulus of sintered alumina, *J. Am. Ceram. Soc.*, **82,** [1] 50-56 (1999)

[7] I. M. Krieger, Rheological studies on sterically stabilized model dispersions of uniform colloidal spheres, *Journal of colloid and interface science*, **113**, [1], 101-113 (1986)

[8] T. Isobe, Dispersion of nano- and submicron-sized Al_2O_3 particles by wet-jet milling method, *Materials Science and Engineering B 148*, 192-195 (2008)

[9] P. Bowen, Colloidal processing and sintering of nanosized transition aluminas, *Powder Technology*, **157,** 100-107 (2005)

[10] H. Y. T. Chen, Adsorption of PAA on the α-Al_2O_3 surface, *J. Am. Ceram. Soc.*, **90**, [6], 1709-1716 (2007)

[11] Kathy Lu , Optimization of a particle suspension for freeze casting, *J. Ceram. Soc.*, **89**, [8], 2459-2465 (2006)

[12] J. Cesarano, Processing of highly concentrated aqueous α-Alumina suspensions stabilized with polyelectrolytes, *J. Am. Ceram. Soc.*, **71**, [12], 1062-1067 (1998)

[13] Studart et al., Rheology of concentrated suspensions containing weakly attractive alumina nanoparticles, *J. American Ceramic Society*, **89**, 2418-2425 (2006)

[14] T. Chartier, Ultrasonic Dispersion of Ceramic Powders, *J. Am. Ceram. Soc.*, **73**, [8], 2552-2554 (1990)

[15] I. Santacruz, Preparation of high solids content nanozirconia suspensions, *J. Ceram. Soc.*, **91**, [2], 398-405 (2008)

[16] M. Aoki, Analysis and modeling of the Ultrasonic Dispersion Technique, *Advanced Ceramic Materials*, **2**, [3A], 209-12 (1987)

[17] K. Sato, Ultrasonic Dispersion of TiO2 Nanoparticles in Aqueous Suspension, *J. Am. Ceram. Soc.*, 1–7 (2008)

[18] R. Rajan, Correlations to predict droplet size in ultrasonic atomisation, *Ultrasonics*, **39**, 235-255 (2001)

[19] F. Iskandar, Control of the morphology of nanostructured particles prepared by spray drying of nanoparticle sol, *Journal of Colloid and Interface Science*, **265**, 296-303 (2003)

[20] to be published

Author Index